# Laboratory Text
# in Introductory Microbiology
# for Health Sciences Students

# Laboratory Text in Introductory Microbiology for Health Sciences Students

*Stephen A. Norrell*
University of Alaska, Anchorage

PRENTICE-HALL, INC., ENGLEWOOD CLIFFS, NEW JERSEY 07632

*Library of Congress Cataloging in Publication Data*

Norrell, Stephen A. (date)
   Laboratory text in introductory microbiology for
health sciences students.

   1. Medical microbiology—Laboratory manuals.
2. Bacteria, Pathogenic—Identification—Laboratory
manuals.   I. Title.
QR63.N58   1985      616'.01      84-24784
ISBN 0-13-521253-7

Editorial/production supervision
   and interior design: Nancy Velthaus
Cover design: Diane Saxe
Manufacturing buyer: John Hall

© 1985 by Prentice-Hall, Inc., Englewood Cliffs, New Jersey 07632

*All rights reserved. No part of this book may be
reproduced, in any form or by any means,
without permission in writing from the publisher.*

Printed in the United States of America

10  9  8  7  6  5  4  3  2  1

ISBN 0-13-521253-7 01

PRENTICE-HALL INTERNATIONAL, INC., *London*
PRENTICE-HALL OF AUSTRALIA PTY. LIMITED, *Sydney*
EDITORA PRENTICE-HALL DO BRASIL, LTDA., *Rio de Janeiro*
PRENTICE-HALL CANADA INC., *Toronto*
PRENTICE-HALL HISPANOAMERICANA, S.A., *Mexico*
PRENTICE-HALL OF INDIA PRIVATE LIMITED, *New Delhi*
PRENTICE-HALL OF JAPAN, INC., *Tokyo*
PRENTICE-HALL OF SOUTHEAST ASIA PTE. LTD., *Singapore*
WHITEHALL BOOKS LIMITED, *Wellington, New Zealand*

# Contents

PREFACE xiii

**1 INTRODUCTORY MICROSCOPY 1**

**2 BACTERIOLOGICAL CULTURE MEDIA 9**

**3 STERILIZATION OF MEDIA AND LABORATORY WARE 14**

**4 QUANTITATIVE METHODS IN BACTERIOLOGY 18**

### Exercises

1. Isolation Techniques and Colonial Morphology of Bacteria  *25*
2. Preparation of Smears, Simple Stains, and Wet Mounts  *37*
3. Differential Staining Procedures  *47*
4. Additional Differential Staining Procedures  *55*
5. Physiological Characteristics of Bacteria—Reactions of Carbohydrates  *65*
6. Physiological Characteristics of Bacteria—Reactions of Proteins  *72*
7. Physiological Characteristics of Bacteria—Miscellaneous Reactions  *77*
8. Growth of Anaerobic Bacteria  *85*
9. Development and Use of Diagnostic Keys  *98*
10. Use of Diagnostic Keys: Bacterial Unknowns  *106*
11. Throat Cultures  *112*

12. Urinary Tract Cultures  *123*
13. Gastrointestinal Tract Cultures  *135*
14. Lactobacillus Activity in Saliva  *147*
15. Antibiotic Sensitivity Testing: The Kirby-Bauer Procedure  *155*
16. Detection of Mutant Strains of Bacteria  *163*
17. Serological Reactions  *172*
18. Assay of Antimicrobial Agents: Disc-Diffusion Methods  *181*
19. Assay of Antimicrobial Agents: Use-Dilution Methods  *189*
20. Ultraviolet Light as a Bacteriocidal Agent  *197*
21. Calibration of the Microscope  *205*
22. Cell Dimensions: Measurement of Cell Size  *213*
23. Bacterial Population Counts: Microscope Counting Methods  *221*
24. Bacterial Population Counts: Viable Cell Counts  *232*
25. Spectrophotometric Methods  *243*
26. Viral Population Counts: Plaque Counting  *256*
27. Phage–Host Cell Interaction
    Part One: Plaque Morphology  *265*
28. Phage–Host Cell Interaction
    Part Two: Host-Cell Specificity  *273*

## Appendices

1. References  *282*
2. Media, Supplies, and Staining Procedures  *283*
3. Chemicals, Prepared Reagents, and Test Procedures  *287*
4. Cultures Used in this Manual  *291*
5. Bacterial Characteristics Forms  *293*
6. The Microscope: Its Use and Routine Care  *316*

# Expanded Contents

PREFACE xiii

## INTRODUCTORY CHAPTERS

**CHAPTER 1  INTRODUCTORY MICROSCOPY  1**

Resolving Power and Numerical Aperture
Bright-Field Microscopy
Dark-Field Microscopy
Fluorescence Microscopy
Phase-Contrast Microscopy
Quantitative Measurements With the Microscope

**CHAPTER 2  BACTERIOLOGICAL CULTURE MEDIA  9**

Buffering and Osmotic Protection
Agar and Broth
Commercial Media
Prepoured Media

**CHAPTER 3  STERILIZATION OF MEDIA AND LABORATORY WARE  14**

Autoclave: Wet Heat
Dry Heat
Sterilization of Heat-Labile Liquids: Filtration
Sterilization of Plastics: Radiation and Gas
Quality Control and Sterility Checks

## CHAPTER 4  QUANTITATIVE METHODS IN BACTERIOLOGY  18
        Quantitative Streak Plates
        Serial Dilutions
        Pour-Plate Counts
        Surface-Plate Counts
        Membrane-Filter Counts
        Most Probable Number Estimations

## EXERCISES

**Exercise 1. Isolation Techniques  25**
        Aseptic Techniques
        Streak Plates
        Pour Plates
        Colonial Morphology
        Laboratory Report Form

**Exercise 2. Preparation of Smears, Simple Stains, and Wet Mounts  37**
        Bacterial Cell Morphology
        Smears and Simple Stains
        Wet Mounts and Motility
        Laboratory Report Form

**Exercise 3. Differential Staining Procedures  47**
        Gram Stain
        Acid-Fast Stain
        Metachromatic-Granule Stain
        Laboratory Report Form

**Exercise 4. Additional Differential Staining Procedures (Demonstration of Bacterial Cytology)  55**
        Negative Stain
        Capsule Stain
        Spore Stain
        Flagella Stain
        Laboratory Report Form

**Exercise 5. Physiological Characteristics of Bacteria  65**
        Part One: Reactions of Carbohydrates
           Fermentation of Sugars
           Methyl Red Test
           Citrate Utilization
           Voges-Proskauer Test
           Hydrolysis of Starch

**Exercise 6. Physiological Characteristics of Bacteria  72**
        Part Two: Reactions of Proteins
           Hydrolysis of Gelatin
           Indole Production
           Hydrogen Sulfide Production
           Urea Hydrolysis
           Nitrate Reduction

EXPANDED CONTENTS  ix

Exercise  7.  **Physiological Characteristics of Bacteria  77**
    Part Three: Miscellaneous Reactions
        Catalase Test
        Oxidase Test
        Coagulase Test
        DNAase Test
        Esculin Hydrolysis
        Motility
    Laboratory Report Form

Exercise  8.  **Growth of Anaerobic Bacteria  85**
    Obligate and Facultative Organisms
    Growth in Thioglycolate Medium
    Isolation of Anaerobic Bacteria
    Laboratory Report Form

Exercise  9.  **Development and Use of Diagnostic Keys  98**
    Adansonian (Numerical) Taxonomy
    Dichotomous Keys
    Profile Analysis
    Phenotypic Profile Analysis
    Laboratory Report Form

Exercise 10.  **Use of Diagnostic Keys: Bacterial Unknowns  106**
    Development of Diagnostic Protocols
    Laboratory Report Form

Exercise 11.  **Throat Cultures  112**
    Gram-Positive and Gram-Negative Cocci
    Blood and Chocolate Agars
    Dichotomous Key of Bacteria in This Exercise
    Laboratory Report Form

Exercise 12.  **Urinary Tract Cultures  123**
    Calibrated Loop Counting
    **Gram-Negative** Cocci
    Dichotomous Key of Bacteria in This Exercise
    Laboratory Report Form

Exercise 13.  **Gastrointestinal Tract Cultures  135**
    Gram-Negative Rods
    Selection for Salmonella and Shigella
    Selection of Lactose Nonfermenters
    Selection of Lactose Fermenters
    Dichotomous Key of Bacteria in This Exercise
    Laboratory Report Form

Exercise 14.  **Lactobacillus Activity in Saliva  147**
    Estimation of Lactic Acid Production by Bacteria in Saliva
    Laboratory Report Form

Exercise 15.  **Antibiotic Sensitivity Testing: The Kirby-Bauer Procedure  155**
    Use of Kirby-Bauer Procedure
    Disk-Diffusion Techniques
    Effect of Concentration on Zone Size
    Laboratory Report Form

**Exercise 16. Detection of Mutant Strains of Bacteria  163**
Mutation Induction by Ultraviolet Light
Selection of Antibiotic Resistant Mutants
Concentration–Independent Mutants
Concentration–Dependent Mutants
Laboratory Report Form

**Exercise 17. Serological Reactions  172**
Agglutination Reactions
Blood Grouping and Typing
Serotyping of Salmonella
Febrile Agglutination Reactions
Laboratory Report Form

**Exercise 18. Assay of Antimicrobial Agents: Disc-Diffusion Methods  181**
Disk-Diffusion Techniques
Bacteriostatic Agents
Effect of Media Components on Effectiveness
Laboratory Report Form

**Exercise 19. Assay of Antimicrobial Agents: Use-Dilution Methods  189**
Use-Dilution Techniques
Bacteriocidal Agents
Effect of Surface on Agent Effectiveness
Laboratory Report Form

**Exercise 20. Ultraviolet Light as a Bacteriocidal Agent  197**
Mutagenic Activity of Ultraviolet Light
Effect of Suspending Medium on Ultraviolet Light
Laboratory Report Form

**Exercise 21. Calibration of the Microscope  205**
Stage and Ocular Micrometers
Determination of Diameter and Area of Field of View
Calibration of the Ocular Micrometer
Laboratory Report Form

**Exercise 22. Cell Dimensions: Measurement of Cell Size  213**
Diameter Measurements: Diameter of Red Blood Cells
Diameter and Length Measurements: Yeast and Bacteria
Additional Calculations for Determining Biomass
    Volume of Red Blood Cells
    Volume of Rods and Cocci
Laboratory Report Form

**Exercise 23. Bacterial Population Counts: Microscope Counting Methods  221**
Use of Counting Chambers: Blood Cell Counts
Use of Breed Smears: Counting of Bacteria in Milk
Laboratory Report Form

**Exercise 24.** **Bacterial Population Counts: Viable Cell Counts** 232
　　　　　　　Serial Dilutions
　　　　　　　Pour-Plate Counts
　　　　　　　Membrane-Filter Techniques
　　　　　　　Laboratory Report Form

**Exercise 25.** **Spectrophotometric Methods** 243
　　　　　　　Relationship between Turbidity and the Density
　　　　　　　　of the Cell Suspension
　　　　　　　Part One: Standard Curves: Absorbance/Plate Count
　　　　　　　Part Two: Population Changes
　　　　　　　Laboratory Report Form

**Exercise 26.** **Viral Population Counts: Plaque Counting** 256
　　　　　　　Plaque Counting
　　　　　　　Agar Overlay Technique
　　　　　　　Laboratory Report Form

**Exercise 27.** **Phage–Host Cell Interaction**
　　　　　　　**Part One: Plaque Morphology** 265
　　　　　　　Coliphage T4 and T4r
　　　　　　　Rapid Lysis Mutants
　　　　　　　Plaque Differentiation
　　　　　　　Laboratory Report Form

**Exercise 28.** **Phage–Host Cell Interaction**
　　　　　　　**Part Two: Host-Cell Specificity** 273
　　　　　　　Coliphage T2 and $\phi$X174
　　　　　　　Serotyping
　　　　　　　Permissive Host Cell
　　　　　　　Laboratory Report Form

## APPENDICES

**Appendix 1.** **References** 282
　　　　　　　Text References
　　　　　　　Resource References

**Appendix 2.** **Media, Supplies, and Staining Procedures** 283
　　　　　　　Media Used in This Manual
　　　　　　　Supplies Needed
　　　　　　　　Individual Supplies
　　　　　　　　General Supplies
　　　　　　　Preparation of Staining Reagents

**Appendix 3.** **Chemicals, Prepared Reagents, and Test Procedures** 287
　　　　　　　Chemical Reagents
　　　　　　　Reagent Preparation and Test Procedures
　　　　　　　Prepared Reagents and Disks
　　　　　　　Antibiotics
　　　　　　　Serological Reagents

xii  EXPANDED CONTENTS

**Appendix 4. Cultures Used in This Manual  291**
    Bacterial Cultures
    Phage Cultures
    Yeast Cultures
    Stock Cultures

**Appendix 5. Bacterial Characteristics Form  293**

**Appendix 6. The Microscope: Its Use and Routine Care  316**
    Routine Care and Cleaning
    Use of the Microscope
    Binocular Microscopes

# Preface

## ABOUT THE MANUAL

This laboratory manual is not, and is not intended to be, a manual for general microbiology. It was written specifically for students in allied health programs, and its contents are limited to those topics appropriate to the allied health profession. It is intended for those students who take microbiology during their freshman or sophomore year and who, typically, have had only one year (or less) of chemistry. More importantly, this manual recognizes that while most allied health professionals do not practice microbiology, they should be familiar with certain aspects of it, such as diagnostic microbiology, clinical samples and sampling, serological procedures, antibiotic sensitivity testing, and so on. Other manuals are available that are more appropriate for general microbiology courses, or for more advanced students.

The manual has several important features. First, it is hoped that it will be found to be reasonably self-contained. Each exercise, in addition to the usual Materials Needed and Methods sections, contains detailed explanations of the salient points being demonstrated or tested. These explanations are, in some cases, textlike, written to reinforce the learning process when used with the assigned textbook, but written in a relaxed yet clear style. The explanations contained in each exercise and in the introductory chapters are meant to supplement and complement, not replace, your assigned text. Each exercise has been referenced to seven popular bacteriology textbooks and to five resource references that the author has found helpful.

In addition to a detailed Laboratory Procedures section, a Laboratory Protocol section is included at the end of each exercise. These protocols may be used as a checklist to walk through each procedure on a first-day and second-day basis. Finally, a separate page for notes is included with each exercise.

The assignments are arranged into related "packages" of material (often offered as separate exercises in other manuals) that could be set up by the average student (or pairs of students) in one laboratory period (of about three hours' length). In several instances, the assignments are arranged so that one member of the pair completes one experiment while the other member completes the other experiment.

It is assumed that in those cases where there is not a scheduled second laboratory period, you will return to the laboratory at your convenience to complete the assigned second-day tasks. Alternately, the reading of your experimental results and class discussion of those results could be provided for by a second scheduled laboratory period in each week.

You will find four introductory chapters at the beginning of the manual. If you read these carefully, you will have a much better understanding of the topics contained in the exercises. Your knowledge of their contents will markedly reduce the amount of class time spent lecturing on basic microscopy, media, sterilization procedures, and quantitative procedures.

Last, the author has tried to use a relaxed style of writing. Learning should be fun, and much of that fun is lost by using a manual that rigidly presents a protocol, "cookbook" fashion without giving an explanation of the "whys and what fors." The procedures are given in protocol format, but only after lengthy discussion of the material in the exercise. Microbiology can be interesting and fun. I hope this manual helps you enjoy the course a little more.

## DOING YOUR PART

Exercises 1 through 17 should be completed by all students. They are, in the opinion of the author, essential and cover those aspects of microbiology most likely to be encountered by health sciences practitioners in their careers. The remaining exercises should be used as time and individual class requirements dictate. They illustrate some additional principles in microbiology, all of which have some important relationship to health sciences.

Exercises 21-25 are essentially quantitative in content, but also address important applications of quantitative techniques, such as blood-cell counting, spectrophotometric techniques, and plate counting and serial dilutions. Some classes may complete all five of these, while others may choose to complete only one or two.

As you proceed through this text, you may notice that the laboratory exercises get progressively more complex and the instructions get somewhat less comprehensive as the semester proceeds. You are expected to become increasingly more competent with each exercise and should be able to figure some things out for yourself.

Laboratory reports are required for all exercises, and report forms are included for each exercise. They require tabulation of the data acquired in the exercise and include several questions based on the experimental procedures and principles demonstrated in the exercise. Each student should complete a laboratory report form for each exercise even if students were allowed to work in pairs. Bacterial characteristics forms are also provided (Appendix 5). One form should be completed for each organism encountered throughout the manual. It may be necessary to look up some of the characteristics, but most of the data needed for the forms will be generated in Exercises 1 through 7 in the manual. The forms should prove useful in helping you get to know your organisms.

## GETTING THE MOST OUT OF EACH EXERCISE

Each exercise is divided into several sections. The first section contains a discussion of the theory and experimental design underlying the procedures used in the exercise. This is followed by a statement of the objectives, then a list of the materials needed to complete the exercise. The specific instructions for completing the laboratory assignment will be found in the Laboratory Procedure section.

References from popular textbooks and reference manuals are listed in the References section. The text references refer you to appropriate sections of your text, while the resource references provide a more detailed, in-depth explanation of the procedures.

At the end of each exercise you will find a section labeled Laboratory Protocol, containing a list of the procedures you should follow to complete the assignment. There are spaces for you to check off each step as you complete it. Finally, each exercise includes a blank page for notes and other information you might wish to record. Laboratory Report Forms are included at the very end of each exercise, where the text of the exercise is readily available for reference.

## LABORATORY SAFETY

Microbiology laboratory exercises are somewhat different from those you might have experienced in most other courses. Microbiology demands the use of live organisms. Most of the experiments in this manual require that you make observations and tests on living and actively growing bacteria. Some of the bacteria are pathogenic and have been associated with laboratory infections. You must, therefore, use caution to prevent accidental infections.

There is, however, another important reason to develop good techniques for handling material that is potentially infectious. It is very likely that over the course of your career, you will be asked to handle infectious material. Syringes, urine samples, clothing—virtually any item that has been used by a patient is potentially infectious. As a health professional, you are obligated to do all you can to prevent spread of any diseases, including those encountered in your professional activity. There are also many instances where the maintenance of sterility is critical (e.g., surgery) and where the introduction of even noninfectious bacteria is a serious circumstance.

You will be introduced to good safety techniques in this course; it is hoped that their use will become habitual. You will almost certainly use those habits for the rest of your professional life, and they will protect you and your patients from accidental infection. Good safety techniques and practices include the following:

1. Wash your hands carefully before and after working in the laboratory. Dispose of the paper toweling in a safe way.
2. Never remove anything from the laboratory that has not been sterilized without the permission of your instructor. This is especially true for bacterial cultures.
3. Spills must be reported immediately. Follow any cleanup instructions carefully and exactly.
4. Never eat, drink, smoke, or place anything in your mouth in the laboratory. Your instructor will give you special instructions for pipetting of liquids containing bacteria.
5. Never use gummed labels that require wetting in the laboratory. Use only labels and tape that are self-sticking and do not require wetting before application.
6. Carefully wash your work area with disinfectant before you begin your work and after you complete it.
7. Be especially careful to develop your aseptic techniques as they are presented to you. Most laboratory infections are the result of poor or lapsed technique.
8. Laboratory aprons, jackets, or coats should be worn at all times. Any clothing that has been contaminated must be sterilized before being removed from the laboratory.

*Stephen A. Norrell*
*Department of Biological Sciences*
*University of Alaska, Anchorage*

# Chapter 1
# Introductory Microscopy

The instrument you will use to examine cellular morphology of bacteria is properly called a *compound light microscope*. In this chapter we will review the microscope, its construction, and its correct use.

You are, of course, familiar with a magnifying glass. The original "microscopes," as described by Leeuwenhoek, were simple devices consisting of a single lens and focusing mechanism. Leeuwenhoek's contribution was to describe techniques for producing lenses with very high magnifications and for using a mechanical device for focusing on very small samples. Later scientists realized that the image of one lens could be further magnified with a second lens. The compound microscope was born. Our modern optical instruments—telescopes, microscopes, and binoculars—are *compound* optical instruments—one lens system magnifies an image produced by the other.

## THE COMPOUND MICROSCOPE

For purposes of discussion we will consider the microscope to consist of three systems: the lens system, the focusing and sample processing system, and the light-source system. Consult the diagram in Figure 1-1 as you read through the following paragraphs.

One of the simplest measures of a microscope's effectiveness is its *resolving power*. Resolving power is an expression of the smallest distance that two objects can be to each other and still be perceived as two distinct objects. *Magnification,* on the other hand, simply measures how many times the lens systems of the microscope increase the apparent size. To understand the difference between resolving power and magnification, you should consider what would happen if you used grainy film to make a highly enlarged photograph—all you would have would be a large blurry picture. A microscope with high magnification but poor resolving power produces a large blurry image—one that is of little value. A good microscope is one that has reasonable magnification and good resolution—and, if a choice has to be made between the two, it is always made in favor of resolution. We will say more about that later.

### The Lens System

The two lens systems found in a compound microscope are the *objective lens,* the one closer to the sample, and the *ocular lens,* the one closer to your eyes. (Examine the upper diagram on Figure 1-1.) The distance between the two, the path through which the light (and images) must pass, is called the *light path*. Some micro-

**2** *INTRODUCTORY MICROSCOPY*

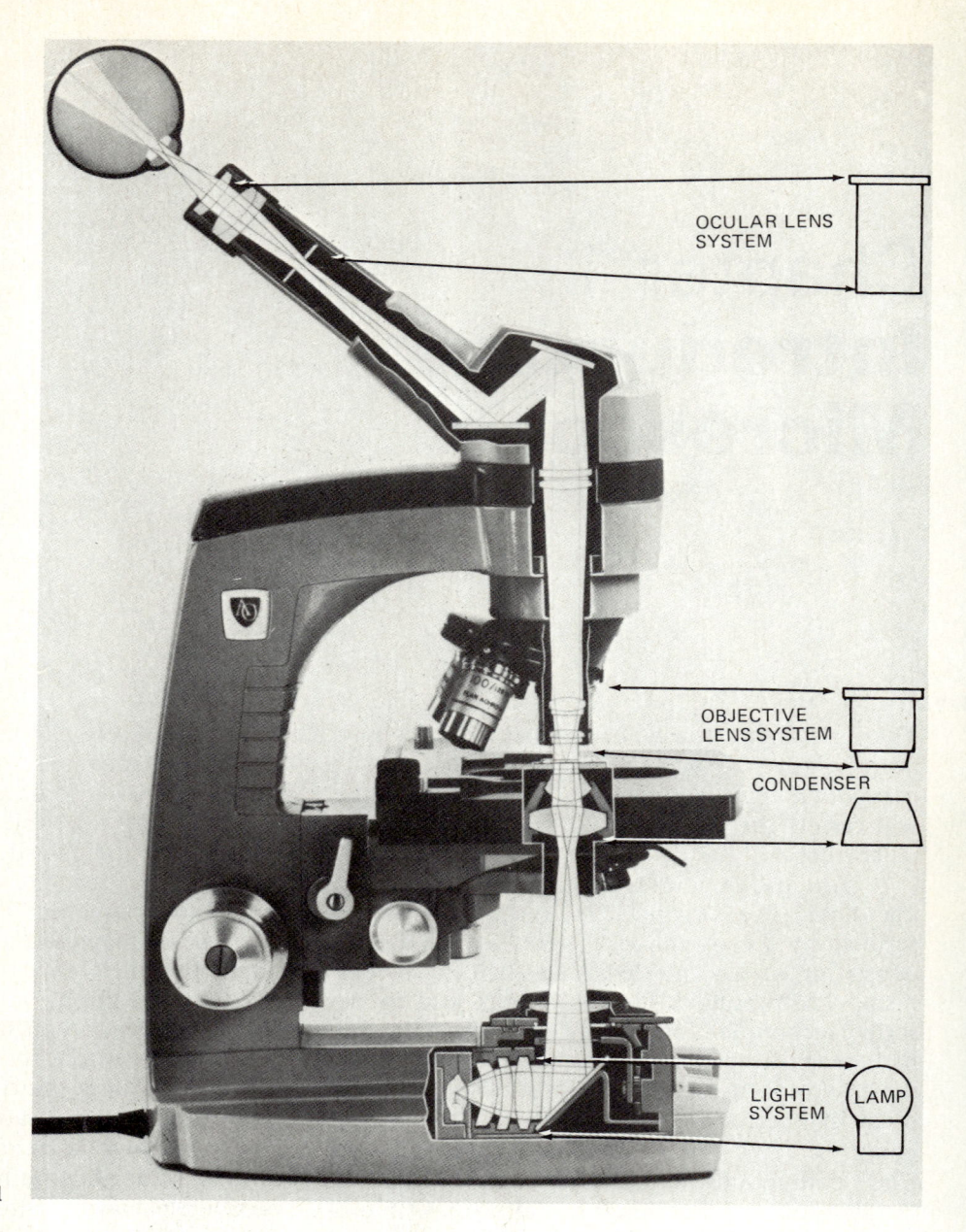

Figure 1-1

scopes, such as those that are inclined, or are binocular (with two ocular lens systems), have prism systems to "bend" the light path or to split it as needed. The prism systems usually do not add substantially to the quality of the microscope, although they do increase the length of the light path somewhat.

Each of these lens systems is quite complex. The reason for this is twofold: First, multiple lenses are cemented together to compensate for aberrations resulting from refraction of light, curvature of field, and color aberrations that occur when a light path is bent (remember your first experience with making "rainbows" with prisms). The second reason has to do with producing a workable focal length. The *focal length* is the distance the lens must be from something to focus on it. Remember how you had to move the magnifying glass closer to an object in order to see it clearly? When you did this, you determined the focal length of the magnifying glass (in a crude but workable way). Remember that the two lenses (ocular and objective) are designed to be used together.

Recall now that the compound microscope uses one lens system to magnify the image produced by the other. It is essential that the distance between the lenses (the light path) be carefully controlled and that the correct lenses be used in the microscope. Even the best lenses will produce poor images if the image produced by one lens does not fall precisely at the focal length of the other lens. The light path of the microscope must be constant and cannot change. If it does, it would not function optimally as a compound microscope.

There is one more important characteristic of objective lenses that we need to mention. In addition to magnification, each lens has a specified numerical aperture. As we shall see, the *numerical aperture,* or *NA,* is very important for maximum resolution. We will discuss it in more detail later, but for now we can think of the NA as a way of describing the shape of the cone of light that should be delivered to the objective lens. If the light cone is too small, the image will appear dark; if it is too large, it will appear glary. In either case, resolution will suffer seriously.

The magnification of a compound microscope is equal to the magnification of the objective lens multiplied by the magnification of the ocular lens. The magnification and the numerical aperture will be printed on the body of the lens. Typically, oil-immersion lenses have a magnification of 100 times and a numerical aperture of 1.25.

In summary, then, the lens system of a compound microscope consists of two lens systems: the objective lens and the ocular lens. They must be precisely located relative to each other. The shape of the light cone entering the objective lens should be controlled to match the numerical aperture of the lens and to provide maximum resolution.

### The Light-Delivery System

The light system in modern microscopes consists of the light source (an electric light) and a focusing system.

The light source may be an ordinary incandescent bulb, although special bulbs can be used if the wavelength of emitted light is important. Bulbs used for general microscopy produce good emissions over most of the visible spectrum. Light sources for special-purpose systems either have lamps that emit only over a narrow part of the visible spectrum or have filter systems to remove the unwanted wavelengths. The wavelength of light used, as we will see later, does affect resolution. As a rule, the shorter the wavelength, the better the resolution.

The focusing system for the light source is called the *substage condenser,* or simply the *condenser.* Condensers are designed to deliver a cone of light whose dimensions are specified. Remember how you could focus the light from the sun into a very small (and very hot) spot with a magnifying glass? The condenser does the same with the light produced by the light source. The difference, however, is that the cone of light produced by the lenses in the condenser must match the numerical aperture of the objective lens. The condenser will have its numerical aperture printed on it and, ideally, should match the numerical aperture of the objective lens. When they are matched, the condenser will be able to deliver the light to the objective lens in a way that will allow maximum resolution.

The actual dimensions of the cone of light entering the objective lens are determined by the lenses and iris diaphragm in the condenser. You can control the diameter of the cone by adjusting the diaphragm. The position of the condenser (how far below the stage it is) determines how closely the cone matches the objective-lens opening. Proper adjustment of the condenser is essential for maximum resolution.

### Focusing System

Modern compound microscopes focus on the subject by moving either the lenses or the stage relative to each other. On some microscopes the stage (the flat platform that holds the slide) moves up and down; on others the body of the microscope moves. In either case, there is always a *coarse* and *fine* focusing adjustment. Adjustment is complete when the subject is in sharp focus. A good microscopist will alternately adjust the focus and the condenser to achieve maximum resolution.

### Resolution

The resolving power, or resolution, of a microscope is expressed as:

$$\text{resolving power} = \frac{0.5 \times \text{wavelength}}{N \times \text{sine of } a}$$

and

$$\text{numerical aperture} = N \times \text{sine of } a$$

Substituting,

$$\text{resolving power} = \frac{0.5 \times \text{wavelength}}{\text{numerical aperture}}$$

Resolving power describes how close two points can be together before they are no longer perceived as two separate points. The smaller the resolving power, the better. If we examine these relationships mathematically, we can see how the components of the system can be used to increase resolving power (making it as small as possible).

Using a short wavelength of light will increase resolution. Many microscopes have provision for inserting a blue filter in the light path. This reduces the average wavelength of light, thereby gaining some resolution. Of course, if the light gets too close to the ultraviolet part of the spectrum, special filters must be used to prevent damage to the eyes. If the wavelength is outside the visible spectrum we wouldn't be able to see it anyway.

In the formula for numerical aperture, the sine of $a$ is determined by the optics of the condenser; $a$ is one half the angle of the cone of light; $N$ is the refractive index of the medium between the specimen and the objective lens. Air has a refractive index of 1.00, while good-quality immersion oil has an index of 1.25. The use of immersion oil has two important effects on resolution: first, it increases the numerical aperture, thereby decreasing the value for resolution (look at the formulas). Secondly, it prevents loss of light due to *refraction*. Refraction is the bending of light as it passes from one transmitting medium to another. (Remember the stick in the water, and how it appeared to bend!) The refractive index of immersion oil (about 1.25) approximates that of the glass in the lens, thereby minimizing loss of light. It is necessary to use immersion oil because it increases the resolution of the instrument *and* because it minimizes light loss due to refraction.

To summarize, we can increase the resolution of the microscope by reducing the wavelength of light used and by increasing the numerical aperture with a good condenser and immersion oil. The need to carefully adjust the condenser should now be obvious.

## DARK-FIELD MICROSCOPY

In the previous section of this chapter we discussed bright-field microscopy. In bright-field microscopy, a solid cone of light passes through the specimen and enters the objective lens. We can see the specimen because it is dyed (stained) and we see the difference in color, or we see it because it is less transparent than the medium it is suspended in; we see it because it blocks light. Bright-field microscopy works best when the specimen can be stained. Examination of unstained specimens (i.e., wet mounts) is possible, but resolution is poor.

There are many instances, however, when the specimen cannot be stained. For example, we might want to examine a living sample, or the staining procedure might drastically alter the structure of the specimen. When this is the case, dark-field microscopy offers a readily available solution.

The condenser that is used for dark-field microscopy (Fig. 1-2) is very similar to that used for bright-field work. It is different in that an opaque disc is inserted in the light path which causes a hollow cone of light to be emitted from the condenser. The light path used for dark-field microscopy, as well as some others, is shown in the lower diagrams of Fig. 1-1. The condenser is then focused so that only the apex of the cone passes through the specimen. The light then continues on and does not enter the objective lens. The only light entering the objective lens is that which is refracted off the specimen. The field of view is dark, with the specimen lit by oblique illumination. This phenomenon is based on the Tyndall effect, which is best observed when you see the dust particles in air as a small beam of sunlight enters a dark room. We see the dust particles (and the specimen on the microscope slide) because of the light bouncing (refracting) off them.

## FLUORESCENCE MICROSCOPY

Fluorescence occurs when something absorbs light of one wavelength and emits it at another. In *fluorescence microscopy,* the specimen is

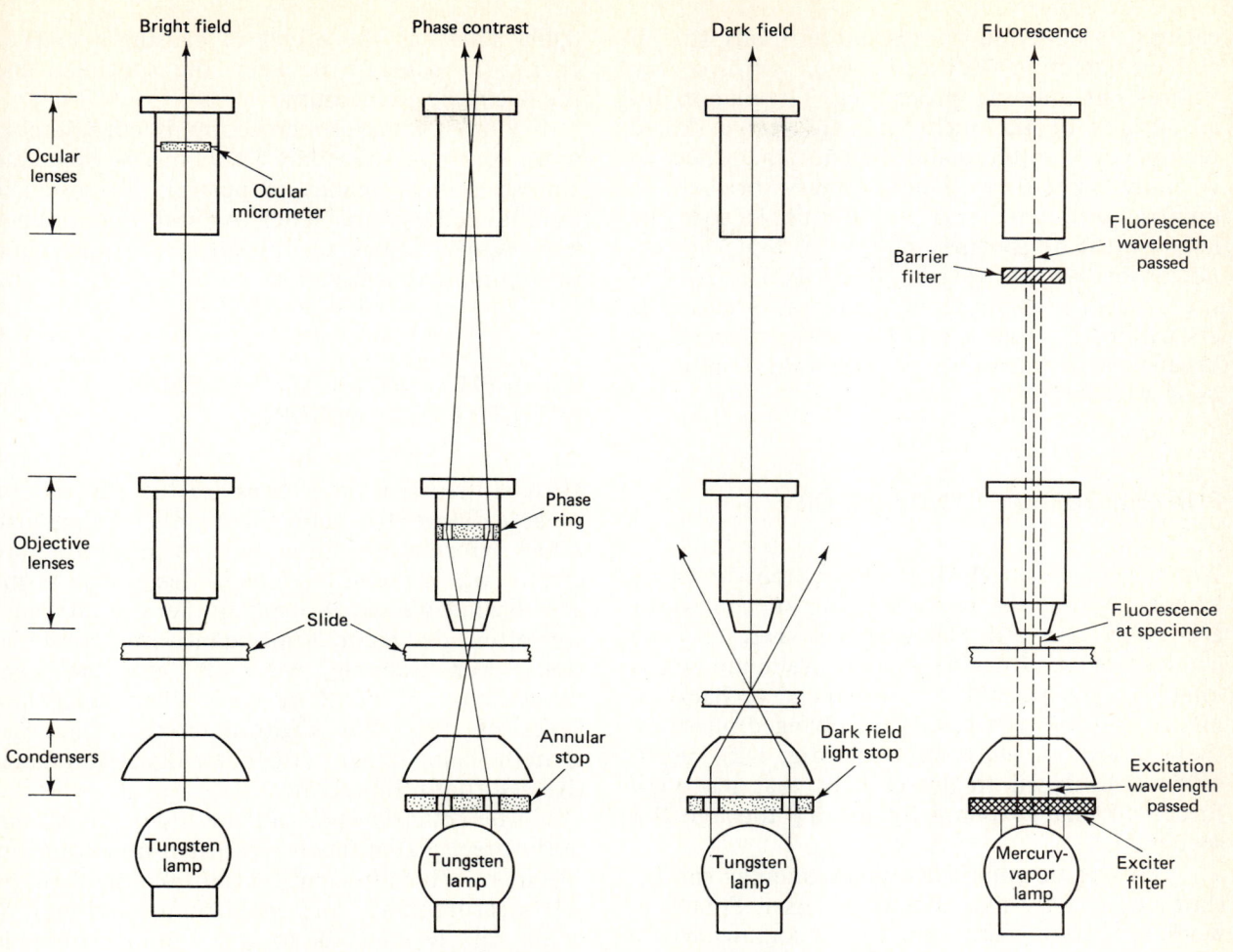

**Figure 1-2** Light-microscopy systems, showing filters and light stops.

stained with a fluorescent dye (sometimes referred to as the *fluorochrome*), exposed to a wavelength of light that will be absorbed by the dye, and viewed through filters that pass only the light emitted by the specimen. In fluorescence microscopy, the field appears dark and the stained specimen appears brightly colored. The wavelength of light that is emitted from the fluorochrome is different from the wavelength of light absorbed by the fluorochrome (i.e., it will be seen as a different color).

The light source in a fluorescent microscope (Fig. 1-2) should be one that emits strongly at the wavelength absorbed by the dye. Some dyes require ultraviolet light (what effect will that have on resolution?). In all cases, filters are used to remove as much of the unwanted wavelengths as possible. Ideally, monochromatic light (at what wavelength?) should be used, but that is usually not practical. The filter that controls the wavelength of light that illuminates the specimen is called the *exciter filter* (Fig. 1-1). It is always located between the light source and the specimen.

A second filter, called a *barrier filter*, is inserted in the light path above the specimen, between the objective and ocular lenses (Fig. 1-1). Its function is to remove all light except that wavelength emitted by the fluorescent dye. The barrier filter is also very important for removing ultraviolet light and is essential for the safe use of fluorescent microscopes. When the exciter filter and barrier filter are properly matched, the field of view should appear very dark (almost black), with the stained specimen appearing brightly lit and colored.

Fluorescence microscopy is an exceptionally sensitive and valuable tool in clinical applications. It is used in fluorescent antibody studies and for detecting acid-fast bacteria with a special fluorescent staining procedure. The major disadvantages of the technique are that a skilled technician is required and the equipment needed is more expensive than ordinary bright-field microscopes. One reason for this is that the lenses must be specially chosen so that they do not absorb at the wavelength of light needed in the procedure. Many types of lenses or condensers cannot be used for fluorescence microscopy because they absorb light at the most commonly used wavelengths.

## PHASE-CONTRAST MICROSCOPY

When light passes through transparent or translucent objects, the characteristics of the waves are often changed. The light is said to have *changed phase*. Imagine a sine wave many cycles long which suddenly jumped ahead or back by ninety degrees! It would have changed phase by that amount. A phase-contrast (Fig. 1-2) microscope is designed to detect light that has been forced to change phase by passing through the specimen.

This is accomplished in a manner similar (but not exactly like) Polaroid lenses. Polaroids work by filtering out light that is somewhat out of phase and appear to greatly reduce glare (refracted and reflected light). A phase-contrast microscope does the opposite (but using the same general principles)—it allows only the changed light to enter the ocular lens. The phase ring in the body of the microscope acts much like the barrier filter does in a fluorescence microscope. It blocks the light that is still in phase and allows only the changed light (out-of-phase light) to pass.

One of the important advantages of phase-contrast microscopy is that the apparent brightness of the changed light is proportional to the square of the amplitude—meaning it will appear four times as bright as it would if a bright-field microscope was used.

Phase-contrast microscopy uses a condenser that arranges the light rays so they are in phase. When the light passes through the specimen, some of it is made to go out of phase. The phase plate, usually located in the specially designed phase-objective lenses (Fig. 1-1), enhances the apparent relative brightness of the out-of-phase light. This has the effect of causing a marked change in contrast between the specimen and the background medium.

Phase-contrast microscopy is considerably more sensitive than dark-field microscopy, and almost always makes it possible to visualize internal cellular organelles. As with fluorescence microscopy, skilled technicians are required and the equipment is expensive.

## QUANTITATIVE MEASUREMENTS WITH THE MICROSCOPE

Most students initially consider the microscope to be a device that helps them gather nonquantitative information about the specimen they are examining. We tend to think in descriptive terms about what we see through the lenses, although we often do make semi-quantitative observations. For example, we might use the term "*small* coccus" or "*long* rod" when describing types of cells. We might describe a bacterial spore as being larger in diameter than the vegetative cell that produces it.

The microscope can be used to obtain fairly precise quantitative information about our specimen. This information can take the form of cell measurements such as diameter and length, or it can be related to the number of cells present in a given sample. When the microscope is properly calibrated, it is possible to use it to measure the size of the cells in the sample and to count the number of cells in the sample.

### The Ocular Micrometer (Figure 1-3)

If the microscope is to be used quantitatively, some means of determining relative lengths must be incorporated into the optical path of the lens system. That is, we must do the equivalent of putting a ruler under the microscope. The most frequently used "ruler," or *ocular micrometer*, is a small piece of glass with a graduated scale etched onto its surface. The graduated scale usually goes from zero to ten, or from zero to one hundred; it is typically further divided into smaller units. These smaller units are exactly one tenth as long as the graduated scale. These smaller units are further subdivided into fifths or tenths. The ocular micrometer (see Fig. 1-2) is placed in the light path, exactly at the focal

INTRODUCTORY MICROSCOPY   7

Ocular micrometer

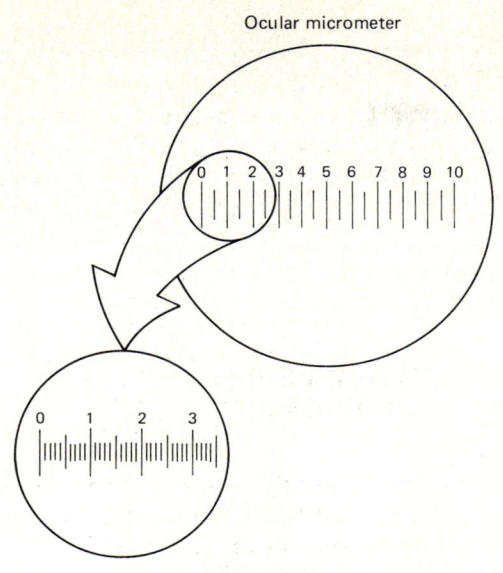

Figure 1-3

Stage micrometer

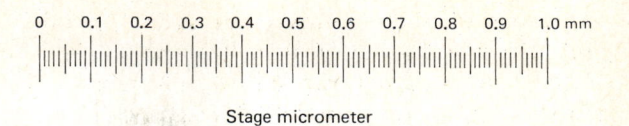

Figure 1-4

length of the ocular lens, where it will be in focus and clearly visible in the field of view. It will appear superimposed over the specimen; if the microscope is carefully focused and the ocular micrometer correctly positioned, both the specimen and the graduated scale will be in sharp focus.

### The Stage Micrometer (Figure 1-4)

Unfortunately, the graduations on the ocular micrometer must be relative length measurements rather than exact ones (like millimeters or micrometers). This is necessary because it is not possible to make every objective lens exactly alike and because our microscopes usually have three or four objective lenses, each giving a different magnification. Remember, also, that the ocular micrometer is positioned at the focal length of the ocular lens and therefore is really being used to determine the dimensions of the image produced by the objective lenses. Consequently, the ocular micrometer must be calibrated for each objective lens, and those calibrations are only valid for the objective lenses and the microscope that they were determined in. The ocular micrometer is calibrated against a scale of known length (like a ruler) that is positioned on the microscope stage and focused like a specimen. This known scale is called a stage micrometer.

A typical stage micrometer has a graduated scale that is of known length. Stage micrometers with scales from 1.0 through 5.0 millimeters are available commercially. For illustrative purposes, we will assume that the scale is exactly one millimeter (1.00 mm), and that it is further subdivided into tenths (1/10; 0.10) and hundredths (1/100; 0.01). Calibration of the ocular micrometer requires that the image of the stage micrometer be superimposed on that of the ocular micrometer. Since the exact dimensions of the stage micrometer are known, the relative dimensions of the ocular micrometer can be easily calculated. Microscope calibration is explained in Exercise 21.

### Applications

After the microscope is properly calibrated and can be used for precise length measurements, several types of determinations are possible, some of which are discussed in this manual. For example, if the diameter of the field of view is known (determined with the use of the stage micrometer), its area can be calculated and reasonably accurate counts of bacteria or other cells can be made. These techniques are discussed in Exercise 23.

Another consequence of the ability to measure lengths accurately is that we can now determine the dimensions of our specimen. The diameter and length of individual cells can be determined by comparing them to the ocular micrometer and using the calibration data to calculate the actual dimensions. Furthermore, since most cells are spherical in at least one plane, once the diameter and length of a cell are known, its surface area and volume can be calculated (Exercise 21). (How would you calculate the area of a circle?)

## REFERENCES

### Text References

1. Atlas, Ronald M., *Microbiology: Fundamentals and Applications*. Chapter 2.

2. Brock, Thomas D., David W. Smith, and Michael T. Madigan, *Biology of Microorganisms,* 4th ed. Chapters 2 and 7. Appendix 5.
3. Jensen, Marcus, *Introduction to Medical Microbiology.* Chapter 6.
4. Nester, Eugene W., et al., *Microbiology,* 3rd ed. Chapter 3.
5. Stanier, R. Y., E. A. Adelberg, and J. Ingraham, *The Microbial World,* 4th ed. Chapters 1 and 2.
6. Stanier, R. Y., et al., *Introduction to the Microbial World.* Chapters 1 and 2.
7. Wistreich, George A., and Max D. Lechtman, *Microbiology,* 4th ed. Chapter 4.

**Resource References**

1. Finegold, Sidney M., and W. J. Martin, *Bailey and Scott's Diagnostic Microbiology,* 6th Ed. Chapter 2.
2. Gerhardt, P., ed., *Manual of Methods for General Bacteriology.* Chapters 1, 3, and 4.
3. Lennette, E. H., ed., *Manual of Clinical Microbiology,* 3rd ed. Chapter 2.

# Chapter 2
# Bacteriological Culture Media

If we are going to be able to grow bacteria, we must provide them, under suitable environmental conditions, with all the nutrients they need. The mixture in which the nutrients are supplied is referred to as the *growth medium* (plural: *media*). The medium also provides the necessary moisture and controls the pH of the environment.

Some media are solidified with a colloidal polysaccharide derived from seaweed (red algae), called *agar*. Agar has no nutrient value and cannot be hydrolyzed (broken down to low molecular weight compounds) by most commonly cultured bacteria. It is simply added to the liquid components of the medium and allowed to solidify. If the medium is used in the liquid form (without the addition of agar), it is called a *broth*. The use of the word *agar* in the name of a medium means that it is used as a solid medium, while the use of the word *broth* means it is used as a liquid medium. The name of the medium also designates the type or origin of the nutrients contained in it. For example, *nutrient agar* is a special kind of solid medium, as is *tryptone broth, MacConkey's* agar, and *glucose yeast extract agar*. Nutrient agar and nutrient broth have the same nutrients, but one (the agar) is solidified, and the other (the broth) is liquid.

## AGAR

Agar, as we have seen, cannot be hydrolyzed by most bacteria and is of no nutrient value to the bacteria. The significance of this is that it can be used as a solidifying agent for bacterial media without changing the components of the medium or without liquefying during the time the bacteria are growing. Agar has several other properties that are also of great importance to the microbiologist. These include·

**1. Remains solid at growth temperatures.** Agar was discovered as the result of a search for a means of separating bacteria into pure cultures. (This is discussed in Exercise 1.) Gelatin, on the other hand, "melts" when the temperature is raised above 25°C and is readily hydrolyzed by many bacteria. When gelatin is used as a solidifying agent, as soon as it liquefies (either because of "melting" or because of bacterial activity) the bacteria become mixed together and can no longer be studied as pure culture populations. Agar does not liquefy until the temperature is raised above about 100°C, and, since it is usually not liquefied by the action of bacteria, can be used to solidify culture media without having to be concerned about the mixing of pure cultures.

**2. Relatively transparent.** Solidified agar is slightly cloudy, but still somewhat transparent. This is an important attribute because it is possible to observe many characteristics of bacterial colonies that would not be apparent if the solidifying agent were opaque. Bacterial colonial morphology is discussed in Exercises 1, 5, and 6.

**3. Physical properties.** Agar is a reversible colloid that can change from the sol (liquefied) phase to the gel (solidified) phase at certain temperatures. The gel to sol (solid to liquid) phase change occurs at about 100°C, but the reverse change, from sol to gel (liquid to solid) occurs at about 45°C. This means that agar can be heated to its liquefaction temperature and then cooled to about 55°C and held in the liquid state until used. While it is liquid, additional components can be added to it and, when necessary, bacteria can be mixed in the agar for separation. If some heat-labile nutrients or bacteria are added to liquid agar when it is at 50°C, the agar can be dispensed into tubes or plates without heat-denaturation of the nutrients or the heat-killing of the bacteria. This, of course, would not be possible if the medium solidified at a higher temperature.

## THE USE OF AGAR

When agar solidifies, it retains its shape until it is liquefied again. There are several ways that solidified agar is used. Remember that the agar must first be liquefied by heating it to about 100°C, then cooling it to about 50°C, and then finally pouring it into either tubes or plates and allowing it to solidify. Solidified agar can be used in any one of the following forms:

*Slants*: A slant is a culture tube (or test tube) in which the medium has solidified while the tube was held in a slanted position. The culture tube is filled about one third (1/3) full with liquefied medium which is allowed to solidify so that the medium has a substantial slant. The slant provides a good growth medium and is an ideal way of maintaining cultures for study.

*Stab*: A *stab*, unlike a slant, is made by allowing the medium to solidify while the culture tube is held in a upright position. The tube is filled between one third and one half (1/3–1/2) full. Stabs are used when reduced oxygen is needed or when it is desirable to observe the effects of growth of bacteria *in* the medium rather than *on* the medium.

*Deeps or pours*: A deep or pour usually contains between 18 and 20 ml of agar medium. They are used to make *pour plates* (Exercise 1). Tubes of medium are considerably easier to store than are plates, and having the medium in a tube makes it considerably easier to handle and manipulate. The medium is liquefied and held in a water bath at about 50–55°C. Components and/or bacteria can be added to it as experimental design requires.

*Plates*: A petri plate is a glass or plastic dish with an overlapping cover. Melted medium is poured into the dish and allowed to solidify. The resulting surface of medium is used to culture bacteria and to separate them into pure cultures by a procedure known as the *streak plate*. In the streak plate the bacteria are spread *over the surface* of the medium, while in the pour plate, the bacteria are *submerged in* the medium.

Figure 2-1

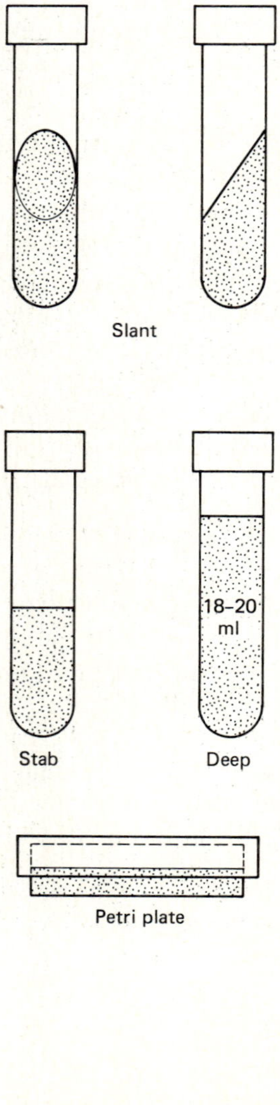

Slant

Stab   Deep

18–20 ml

Petri plate

## Summary

Agar is a reversible colloid that becomes liquid at about 100°C, but does not solidify until the temperature is lowered to about 45°C. It is a polysaccharide derived from seaweed (red algae) that is resistant to hydrolysis by most bacteria. When solidified, agar is semi-transparent, allowing for good visual examination of bacterial colonies. It is added to liquid media to solidify them and may be used as agar slants, stabs, deeps, and/or plates. When it is added to a liquid medium it usually does not change the physiological environment, except for those conditions associated with having to grow *on* a solid medium instead of *in* a liquid medium.

## BROTH

The liquid phase of any culture medium must contain all of the nutrients needed for the growth of the bacteria. Procaryotic cells, like plants, require that their nutrients be dissolved in an aqueous solution. Consult the Text and Resource References at the end of this chapter for a more detailed discussion of the nutritional components of a growth medium.

One of the most important functions of the medium, in addition to nutrient supply, is pH control. Bacteria are usually restricted to a fairly narrow range of pH, although those that do tolerate changes almost always tolerate increased acidity (lower pH) better than increased alkalinity (higher pH). As you know, a mixture of chemicals that is used to control pH is called a *buffer*. The major buffering component of bacteriological media is the proteins and amino acids found in almost all media. Some media, especially those that do not contain large amounts of proteins, may have inorganic buffers added to them. The buffering system in any bacteriological medium is often as important as the nutrients themselves. The control of pH is somewhat complicated by the fact that many bacteria produce organic acids as normal metabolic waste products. The medium must be designed to buffer these acids and to keep the pH within the tolerance range of the bacteria.

The liquid phase of the medium is also used to control the osmotic environment. Bacteria, as a rule, grow best in slightly hypotonic solutions. A *hypotonic solution* is one where the total concentration of solutes is slightly lower than the total concentration of solutes in the cytoplasm of the bacteria. Like all other cells, bacteria cannot tolerate drastic changes in solute concentration. You cannot simply continue to add nutrients or salts to the liquid phase—if you do, you will end up with a medium that cannot support the growth of any bacteria. Remember, the addition of any soluble component to a medium will change its osmotic pressure.

Finally, the liquid phase often contains chemicals that change color under certain environmental conditions. One of the most commonly used examples of this type of compound is a *pH indicator,* or simply an *indicator.* Three commonly used indicators are discussed in Exercise 5. For now, you only need to remember that these chemicals change color when the pH changes. When they are added to a liquid or solid medium, the microbiologist can tell whether or not the pH of the medium has changed by simply looking at it. Other chemicals, including certain dyes, can be used to measure the amount of oxygen dissolved in the medium. These dyes usually turn colorless when there is little or no oxygen dissolved in the medium.

The liquid phase of a culture medium is, as you can see, a very complicated mixture. It must contain all necessary nutrients, some sort of buffering system, and sometimes a chemical indicator. The medium must contain all of these dissolved elements and still be within fairly narrow osmotic ranges. As you can now imagine, the development of a nutrient medium is not a simple task. The procedures for making media must be carefully followed, with special attention given to the amounts of chemicals used. The formula of an established medium should not be changed without careful consideration of what effects these changes will have on the medium's properties.

## COMMERCIAL MEDIA

Fortunately, commercially prepared media are widely available. These products are made in such a way that the microbiologist only has to add a certain amount of the product to a known volume of water to reconstitute the medium. After it is reconstituted, it only needs to be dispensed and sterilized (we will discuss sterilization techniques in the next chapter) before it is used.

Commercial media will have all necessary information printed on the label. This should include a list of the nutrients, buffering compounds, and indicators, giving the amount of each chemical per liter. The label will also tell you the final pH of the medium and will give you instructions for its reconstitution and sterilization. You should take the time to read several of these labels and to compare them with each other. For example, compare the labels for nutrient agar, nutrient gelatin, and nutrient broth! How are they the same? How are they different?

Although commercially produced media are usually of very good quality and are subjected to very rigorous quality controls, it should never be assumed that one bottle of medium will produce exactly the same results as another. Often, subtle differences in color or composition are observed, and although these subtle differences do not significantly alter the results, they may change the appearance just enough to require some modification in the interpretation of results. A good microbiology laboratory will carefully compare a new lot of medium with the old lot so that any such changes can be noted.

One of the most commonly encountered problems that occur with a commercially prepared medium are those associated with the quality of the water used to reconstitute the medium. Although most media have some built-in buffering capacity, the water used to reconstitute the medium may be sufficiently acid or alkaline (usually acid) to use up some of that capacity. This may seriously affect the results obtained with media that require a pH change for the interpretation of results. The pH of all media should be checked before autoclaving and adjusted if necessary.

Special care must be exercised with a medium that is used for stock cultures. The holding of stock cultures for long periods of time on the same medium (in contrast to the 48 hours of a typical experiment) means that the bacteria may be exposed to harmful compounds in the medium for the entire length of time they are held. It is especially important to ensure that good-quality distilled water is used to reconstitute the medium used for stock cultures. Laboratories that do not have a reliable supply of such water should consider using bottled, triple-distilled water to prepare their stock-culture medium. The cost of the water, which is readily available at most drug stores and is always available at hospital-supply businesses, is minimal compared to the cost and inconvenience of losing your stock cultures.

## PREPOURED MEDIA

Virtually all the media used in this manual are available in prepoured, sterile form, either in plates or tubes. Generally, prepoured media are quite expensive and their routine use is usually not warranted. However, prepoured media have certain advantages that make their use desirable in some circumstances. First, the quality of the media, when purchased from reliable firms, is usually excellent. Second, many small laboratories are not equipped to prepare certain types of specialized agars and broths that require filtering or the addition of natural products, such as red blood cells, serum, and heat-labile compounds. Lastly, in those instances where microbiology is only one course in a biology program (in contrast to a microbiology department), and where special media are prepared only infrequently, it would be difficult, if not impossible, to prepare media of a quality equal to what is available in prepoured form.

Commercially available dried media should be used whenever possible. Microbiology programs that do not have well-equipped and adequately staffed preparation facilities should consider using prepoured media whenever the preparation of the media requires more than the usual autoclaving or filtering. A few examples of prepoured media:

- Blood agar, with or without additives
    Blood tellurite medium
    Blood–PEA medium
- Chocolate agar, with or without additives
    Martin-Lewis or
    Thayer-Martin medium
- Biplate combinations
    Blood agar/MacConkey's
    Blood agar/chocolate agar
    Chocolate agar/Martin-Lewis agar
- Serum-containing media
    Lowenstein-Jensen medium (for acid-fast organisms)
    Bile-esculin medium (with serum)
    Loeffler's serum medium (for *C. diphtheriae*)

## REFERENCES

### Text References

1. Atlas, Ronald M., *Microbiology: Fundamentals and Applications.* Chapters 2 and 10.
2. Brock, Thomas D., David W. Smith, and Michael T. Madigan, *Biology of Microorganisms,* 4th ed. Chapters 1, 5, 7, and 8.
3. Jensen, Marcus, *Introduction to Medical Microbiology.* Chapters 6 and 7.
4. Nester, Eugene W., et al., *Microbiology,* 3rd ed. Chapter 4.
5. Stanier, R. Y., E. A. Adelberg, and J. Ingraham, *The Microbial World,* 4th ed. Chapters 1 and 2.
6. Stanier, R. Y., et al., *Introduction to the Microbial World.* Chapters 1 and 2.
7. Wistreich, George A., and Max D. Lechtman, *Microbiology,* 4th ed. Chapter 6.

### Resource References

1. Finegold, Sidney M., and W. J. Martin, *Bailey and Scott's Diagnostic Microbiology,* 6th ed. Chapter 42.
2. Gerhardt, P., ed., *Manual of Methods for General Bacteriology.* Chapters 6 through 10.
3. Lennette, E. H., ed., *Manual of Clinical Microbiology,* 3rd ed. Chapter 97.
4. McGonagle, L. A., *Procedures for Diagnostic Microbiology.* Consult specific topical references.

# Chapter 3
# Sterilization of Media and Laboratory Ware

Bacteria are everywhere; virtually everything that is exposed to the air for even a short period of time should be assumed to be contaminated. All bacteriological media as well as the instruments used to handle bacteria must be sterilized to render them free of contamination. A microbiologist defines something as *sterile* when it is completely free of all living organisms, including bacteria and viruses. The sterilization of media and surfaces requires carefully defined procedures that vary somewhat according to the nature of the materials being sterilized.

Sterilization of heat-stable substances is easily accomplished by heating them to temperatures that will kill bacteria and viruses. Heat-labile substances such as proteins and other medium components must be sterilized by filtration. Solid, heat-labile instruments and culture vessels, such as plastic ware, must be sterilized by the use of toxic gases or lethal levels of radiation.

## BACTERIAL ENDOSPORES AND HEAT RESISTANCE

The bacterial *endospore* is a heat-resistant structure that protects endospore-forming bacteria from the effects of normally lethal temperatures or from many types of disinfectants and harmful environmental conditions. For example, endospores are not affected by prolonged boiling. As a result of this high resistance to heat killing, suspensions of bacterial endospores have become the accepted standard for monitoring the effectiveness of most sterilization procedures. In fact, commercially prepared suspensions of the endospores of *Bacillus subtilis* or *Bacillus sterothermophilus* are often used to test the adequacy of sterilization procedures in hospitals. The objective of heat-sterilization procedures is to kill bacterial endospores, because the endospores are the most heat-resistant of all bacterial structures.

## HEAT STERILIZATION: WET HEAT

The accepted temperature for killing of spores is 121.5°C of wet heat. Since this temperature is above the boiling temperature of water, wet-heat sterilization must be completed in a pressurized container. Containers designed for this purpose are called *autoclaves*; they are able to maintain 121.5°C for as long a time as is needed. At sea level, air-free steam at a pressure of 15 pounds per square inch (psi), will be at 121.5°C. When the autoclave is operated at high elevations, appropriate adjustments in the autoclave controls or procedures are needed. Never forget that the temperature is the critical measure of an autoclave's effectiveness; pressure is merely a convenient way of monitoring an autoclave. The material in the autoclave must be heated to 121.5°C regardless of the pressure needed to reach that temperature.

It is also important to remember that the relationship between pressure and temperature (121.5°C at 15 psi, at sea level) applies only to air-free steam. If all the air is not removed before the pressure vessel is sealed, the required temperature cannot be attained. Most modern autoclaves have air-venting valves that are designed to close only after they have been heated to predetermined temperatures by the escaping steam. These valves are usually quite reliable, but they have been known to malfunction. Any good autoclave will have a temperature gauge as well as a pressure gauge for monitoring the autoclave's performance.

When it is necessary to sterilize liquids, the volume of the liquid must be considered. Heat transfer takes time, and effective sterilization of liquid requires that every part of the liquid be brought up to 121.5°C. When the liquid is contained in test tubes, heat transfer into the center of the tube requires between 10 and 15 minutes. Larger volumes, such as those commonly seen in I.V. bottles or large flasks, require proportionately longer times. As a general rule, sterilization of most laboratory media and instruments calls for autoclaving at 15 psi for 15 minutes. Furthermore, it is essential that the time (15 minutes) be for a consecutive and uninterrupted time span. You cannot autoclave something for ten minutes now and five more minutes later. Thus, ordinary laboratory media and instruments are routinely sterilized by autoclaving for 15 minutes at 121.5°C.

Most physicians' offices have small autoclaves that are manually operated (large autoclaves in hospitals and laboratories are automatic). They should be carefully monitored and tested from time to time (as are the larger, automatic machines). The two critical measures of an autoclave's performance are the temperature (not pressure), and the time that the material is held at that temperature. The "15 minutes/15 pounds" rule is generally adequate unless you are trying to sterilize large volumes of fluids or are at a high elevation.

## HEAT STERILIZATION: DRY HEAT

When it is not convenient or desirable to use wet heat, dry heat may be used. Wrapped instruments, glassware, metalware, and similar non-liquid materials are often sterilized with dry heat. This method of sterilization requires that the instruments and material be heated to 180°C for three hours. As was the case with wet heat, the material must be held at that temperature for the full three hours. Dry-heat sterilization has become somewhat less common than in the past because of the availability of presterilized and disposable plastic ware (pipettes and petri dishes). Nevertheless, glassware and most types of metalware that must be kept dry can be sterilized with dry heat.

## STERILIZATION BY FILTRATION

Many components of media and virtually all drugs, including antibiotics, are heat-labile and cannot not be heated to a high enough temperature to sterilize them, as they would be destroyed by heat. These heat-labile compounds must be sterilized by *filtration*.

Solutions containing such heat-labile compounds are passed through a filter that traps all particles that exceed the pore diameter of the filter. All microbiological filters are rated according to the mean pore diameter, indicating that the filter may be expected to retain all particles with a larger diameter. Bacteriological-grade filters have a pore size that does not exceed 0.45 $\mu$m ($\mu$). Bacteria with diameters larger than 0.45 $\mu$m are retained by the filter. (Some bacteria, particularly certain spirochetes and mycoplasmas, are able to pass through such filters.) Any fluid passing through such a filter will be free of most bacteria. Of course, everything "downstream" from the filter must have been sterilized beforehand to prevent recontamination of the fluids.

Bacteriological-grade filters do not remove viruses. They are smaller than 0.45 $\mu$m and freely pass through the filter. When the fluid is to be used for bacteriological media, the presence of viruses is of little concern. However, if the fluid is to be used for the culturing of viruses, or if drugs and medicines are being sterilized, the viruses must be removed. This may be accomplished by using a filter with a smaller pore diameter. Depending upon circumstances, filters with pore diameters of between 0.1 and 0.22 $\mu$m are used to remove some larger viruses, but would not reliably remove smaller viruses such as the polio or hepatitis viruses from the solution.

Once the heat-labile fluid is filtered and stored in presterilized containers, it must be handled in such a way that recontamination will not occur. Aseptic techniques (what does that mean?) are discussed in Exercise 1. The filter-sterilized fluid can then be added to media, including melted agar (at less than 55°C), as needed.

Although most naturally derived medium additives, such as serum and body fluids, can be sterilized by filtration, whole blood or red blood cells cannot be. Blood from a healthy animal (including human) is sterile and is often added directly to a medium. Alternately, the cells can be separated from the other components of blood, washed, and suspended in sterile saline. Everything that comes in contact with the blood after it is removed from the animal must be sterile. If the blood does become contaminated, it cannot be resterilized and must be discarded (why can't it be filtered?).

## STERILIZATION OF PLASTICS

The plastic ware used in the laboratory or hospital cannot, of course, be heated. It must be sterilized by either exposing it to a toxic gas, usually ethylene oxide, or by exposing it to levels of radiation that are lethal for vegetative bacterial cells and endospores.

If gas sterilization is used, the plastic ware is placed in a large, autoclavelike chamber that is sealed and then filled with the toxic gas. When radiation is used, the plastic ware is exposed to a highly radioactive source for a sufficient period of time. Consult the Text and Resource References for more detailed discussions of gas and radiation sterilization.

The plastic ware will remain sterile only as long as its wrapper remains sealed; it should be considered contaminated if the wrapper and/or seal has been broken in any way.

## WHEN MEDIA MUST BE STERILIZED

All freshly prepared culture media must be sterilized (autoclaved or filtered) before use, within a short time after they have been reconstituted. Many bacteria can reproduce in thirty minutes or less; if a medium is allowed to remain unsterilized for even a few hours, significant growth can occur. A good rule is to sterilize freshly made medium within one hour.

All culture media must be sterilized after they are used. This is necessary to prevent contamination of the environment with the bacteria being studied in the laboratory. This is an especially important consideration for laboratories where pathogenic bacteria are being studied, such as those associated with hospitals, clinics, and universities. In fact, some workers insist that everything coming out of these microbiology laboratories should be sterilized—even the laboratory aprons and coats worn by the technicians.

## QUALITY CONTROL AND STERILITY CHECKS

Every time a tube, plate, flask, or other container of sterile medium is opened, there is possibility of contamination. For example, agar medium for plates (see Chapter 2) must be sterilized before it is poured into the plates. When the flask and/or plates are opened, there is some possibility of contamination. Although aseptic techniques (see Exercise 1) minimize the probability of contamination, they cannot eliminate it. Good quality-control procedures are needed to ensure that plates (and anything else) have not been contaminated. The simplest way to do this is to preincubate the medium overnight before using it. Many good laboratories will use a media-making protocol that calls for overnight incubation of all plates and any other container of medium that have to be opened.

Many laboratories also routinely monitor the performance of their autoclaves. The most common way of doing this is to autoclave bacterial endospores and then attempt to grow the bacteria after autoclaving. Commercially produced kits are available either as saturated filter-paper wicks or small vials containing suspensions of spores. After the spores are autoclaved, they are transferred to a nutrient medium to determine if they are still viable. This type of routine quality control should be a part of every laboratory's protocol. It is especially important when the autoclave is used to sterilize things other than culture media (which quickly show the obvious effects of contamination).

## REFERENCES

### Text References

1. Atlas, Ronald M., *Microbiology: Fundamentals and Applications.* Chapter 10.
2. Brock, Thomas D., David W. Smith, and Michael T. Madigan, *Biology of Microorganisms,* 4th ed. Chapters 7 and 8.
3. Jensen, Marcus, *Introduction to Medical Microbiology.* Chapter 8.
4. Nester, Eugene W., et al., *Microbiology,* 3rd ed. Chapter 6.
5. Stanier, R. Y., E. A. Adelberg, and J. Ingraham, *The Microbial World,* 4th ed. Chapter 2.
6. Stanier, R. Y., et al., *Introduction to the Microbial World.* Chapter 2.
7. Wistreich, George A., and Max D. Lechtman, *Microbiology,* 4th ed. Chapter 18.

### Resource References

1. Gerhardt, P., ed., *Manual of Methods for General Bacteriology.* Chapter 23.
2. Lennette, E. H., ed., *Manual of Clinical Microbiology,* 3rd ed. Chapter 95.

# Chapter 4
# Quantitative Methods in Bacteriology

It is often necessary to estimate the number of bacteria in a sample or to determine the concentration of antibodies in a patient's blood. For example, a primary isolation report is often given as "2+ beta *Streptococcus*" or as "3+ lactose-fermenting gram negative rods"; a urine culture may be reported as the number of bacteria per milliliter of urine, typically as "9,750 bacteria/ml" or as ">100,000 bacteria/ml"; and serum is frequently titrated for antibody content (e.g., "anti-streptolysin-O titre"). Quantitative techniques are also used extensively in public health and environmental health microbiology to determine the number and kind of bacteria in foods, water, sewage, and even air.

All suspensions of bacteria contain living and dead cells. The viable cells can be counted by cultural techniques, such as those explained in this chapter and in Exercise 24. The nonviable (dead) cells must be counted by direct counting methods, such as those covered in Exercises 23 and 25.

There are several quantitative techniques that are used regularly in clinical microbiology and others that are more commonly used in other laboratories. Each has its advantages and each can provide useful information when properly interpreted. The precision of each method varies considerably, ranging from relatively rough estimates to relatively precise counts. Some of the commonly used techniques are:

*Quantitative streak plate*: It is possible to obtain rough estimates of the number of bacteria present in a sample by noting the number and position of colonies deposited from a single streak of a loop or needle. The four-quadrant streak plate is discussed in Exercise 1, and the calibrated-loop streak plate is discussed in Exercise 12.

*Serial dilution*: The concentration of a chemical or an antibody in a sample may be accurately determined by carefully diluting the sample and then testing each dilution for the presence of that substance. The sample is titrated and the endpoint determined. Serial dilutions are used in several exercises in this manual.

*Plate counts*: If a small amount of each dilution (sometimes called an *aliquot*) is transferred to melted agar, and the agar then poured into a petri plate, the bacteria in the sample will grow into visible colonies after suitable incubation (pour-plate count). They can then be easily counted. Alternately, a small amount of the sample can be spread over the surface of agar (surface-plate count), or the bacteria can be trapped on a membrane filter and then allowed to produce visible colonies when the filter is placed over nutrient medium (membrane-filter count). Viruses may be counted using techniques similar to those used for pour-plate counting. A typical plate-count procedure usually is a combination of a serial dilution and the use of a suitable growth medium to detect colonies or plaques in each of the dilutions. Plate-counting techniques are used in all exercises where the number of viable bacteria or viruses must be determined.

*Most probable numbers (MPN)*: A quantitative technique that is widely used in nonclinical applications is the most probable number procedure. It provides the environmental and public health microbiologist with a way of rapidly estimating the most probable number of bacteria in 100 ml of sample. This procedure is briefly discussed later in this chapter.

## SOLUTIONS, SUSPENSIONS, AND DILUTIONS

A word or two on terminology. A solution results when one chemical (the *solute*) is dissolved in another, usually liquid, chemical (the *solvent*). A good example would be a solution of sodium chloride (salt) in water. The salt, being soluble in water, is the solute, while the water is the solvent. The salt is truly soluble in water. A *suspension* results when one of the components is not soluble in the liquid. If you added sand or dirt to water, you would have a suspension of dirt or sand in water. Usually, but not always, solutions are clear while suspensions are turbid.

When you make either a solution or a suspension, you should add the solute to a portion of the solvent, then after it has dissolved (or evenly suspended), bring the mixture up to full volume by adding additional solvent. For example, if you wanted the salt to have a concentration of 1.0 g/100 ml, you would dissolve 1.0 g of salt in 100 ml of water. The concentration would be expressed as 0.01 g/ml or as a 1.0% solution. Similarly, a 1.0% suspension of sand could be made by adding 1.0 g of sand to 100 ml of water. In practice, you would first dissolve the salt (the solute) in about 75 ml of water (the solvent), then bring the volume of the mixture up to 100 ml with fresh solvent.

It is often necessary to further dilute suspensions and solutions. The dilutions are expressed as relative volumes, such as 1 to 10 or 1 to 100 (written as 1:10 or 1:100). Dilutions are made by using pure diluent to change the volume of the sample according to the relative volumes desired. For example, if you wanted to make a 1:10 dilution, you would change the volume from 1.0 ml to 10.0 ml by adding 1.0 ml of the suspension or solution to 9.0 ml of diluent. Using the salt solution as an example, if you transferred 1 ml of the 1% solution to 9 ml of water, you would have made a 1:10 dilution of the original solution. Also, the concentration of the salt would have been reduced by a factor of 10.

Most bacterial and viral samples can be treated as suspensions. If the sample is liquid, we determine the number of bacteria or viruses suspended in the liquid, diluting it as needed. If it is solid, you must first make a suspension (typically, 10% in water) by suspending the sample in a liquid. When these samples are counted, the concentration of the initial suspension must be calculated into the dilution factor. For example, a 1.0% suspension has already been diluted 1:100.

Serological samples are, on the other hand, treated as solutions where the serum or blood is considered to be the undiluted sample. Dilutions are calculated as with suspensions, but the results are reported as titre (discussed in a later section).

## SERIAL DILUTIONS

One of the most important quantitative techniques routinely used in clinical microbiology laboratories is the serial dilution procedure. When a sample has been diluted out, each of the dilutions can be tested to determine which of them give positive test results. Alternatively, the number of bacteria or viruses in each of the diluted samples can be counted after they grow on a suitable medium.

When the amount of antibody in a patient's blood is measured, the report usually indicates the *titre* of the antibody. For example, you might see a report that one of your patients has "an anti-streptolysin-O titre of 512." This means that if you dilute the serum serially, the last dilution that showed the presence of the antibody was the 1:512 dilution. Similarly, a titre of 256 means that the last tube that showed activity was the one that contained the 1:256 dilution. The titre, then, is the reciprocal of the highest dilution that still showed positive activity. Instead of reporting the "the 1:1024 dilution was the endpoint," you may simply report a "titre of 1024."

When the number of bacteria or viruses in a sample must be counted, the report usually indicates the number of bacteria or viruses in each milliliter of the original sample. The report might read "15,000 bacteria/ml" or "150,000 phage/ml." If very large numbers of bacteria are encountered, the logarithmic number is often used instead of the arithmetic number. This, of course, does not change the results at all, but it does make things somewhat more convenient. The actual number of bacteria or viruses is determined by multiplying the number of colonies (or plaques) observed on the plate times the dilution (assuming 1.0 ml aliquots were plated out).

number of bacteria/ml =

number of colonies × dilution

A serial dilution is typically a stepwise dilution sequence made by transferring aliquots from one tube of diluent to another. The aliquots are usually the same volume throughout the sequence. If a 1:10 dilution sequence is used, the process is referred to as a *tenfold serial dilution* (or *twofold*, or *fivefold*). The number (ten, two, or five) is the dilution accomplished at each step, a tenfold serial dilution being a series of 1:10 dilutions. The following table has some examples.

TABLE 4-1. EXAMPLES OF SERIAL DILUTIONS

| Serial | Dilution | Volumes Used |
|---|---|---|
|  |  | Aliquot transferred into |
| Tenfold | 1:10 | 1.0 ml into 9.0 ml |
| Fivefold | 1:5 | 1.0 ml into 4.0 ml |
| Twofold | 1:2 | 1.0 ml into 1.0 ml |

Since a serial dilution is a series of identical, stepwise dilutions, the final dilution, or the dilution of any tube in the sequence, can be easily determined. The actual dilution sequences are arithmetic or logarithmic progressions, the interval or "step" being determined by the dilution used at each step. Diagrams of tenfold serial dilutions are shown in Exercises 24 and 26. A twofold serial dilution is shown in Exercise 25.

As you can see from the data in the table, the highest dilution is always in the last tube, while the lowest dilution is always in the first tube. Using zero (0) to indicate the undiluted sample is sometimes helpful in keeping track of the dilution steps. When bacterial or viral suspensions with very large numbers of organisms are being diluted, the undiluted sample is frequently not counted. When antibody titre in serum is to be determined, or when the expected numbers are very low, the undiluted sample often is counted or assayed.

If you were measuring the anti-streptolysin-O titre of a patient and the last positive reaction was observed in tube #7 of the twofold series (shown in Table 4-2), the titre would be 64. (Tube #7 is a 1:64 dilution; titre is 64.)

TABLE 4-2. EXAMPLES OF TWOFOLD AND TENFOLD SERIAL DILUTIONS, SHOWING TITRES FOR EACH TUBE IN THE SERIES

| Twofold Dilution ||| Tenfold Dilution |||
|---|---|---|---|---|---|
| Tube No. | Dilution | Titre | Tube No. | Dilution | Log No. |
| 1. | Undiluted | Undiluted | Undiluted sample not counted |||
| 2. | 1:2 | 2 | 1 | 1:10 | −1 |
| 3. | 1:4 | 4 | 2 | 1:100 | −2 |
| 4. | 1:8 | 8 | 3 | 1:1000 | −3 |
| 5. | 1:16 | 16 | 4 | 1:10,000 | −4 |
| 6. | 1:32 | 32 | 5 | 1:100,000 | −5 |
| 7. | 1:64 | 64 | 6 | 1:1 million | −6 |
| 8. | 1:128 | 128 | 7 | 1:10 million | −7 |

Additional tubes, as needed to complete the assay

Similarly, if you were using the tenfold series to count bacterial viruses and you counted 123 plaques on the plate produced from the suspension in tube #4, the number of viruses in the original sample would be 1,230,000 viruses/ml (123 × 10,000 [tube #4] = 1,230,000).

When very large numbers of bacteria are expected, the dilution protocol may be varied by using both 1:100 and 1:10 dilutions in the same sequence. For example, if you needed to determine the total number of bacteria in water that has been contaminated with sewage, you might use the scheme shown in Table 4-3.

**TABLE 4-3. SERIAL-DILUTION PROTOCOL FOR HEAVILY CONTAMINATED SAMPLES**

| Tube or Bottle Number | Step Dilution | Final Dilution | Log (10) of Dilution |
|---|---|---|---|
| (The undiluted sample is not counted.) | | | |
| 1. | 1:100 | 1:100 | −2 |
| 2. | 1:100 | 1:10,000 | −4 |
| 3. | 1:10 | 1:100,000 | −5 |
| 4. | 1:10 | 1:1 million | −6 |
| 5. | 1:10 | 1:10 million | −7 |
| 6. | 1:10 | 1:100 million | −8 |

As you can see from the data in Table 4-3, when the bacterial or viral count is expected to be very high, you may use one or more 1:100 dilutions initially. This will allow for rapid dilution and does save both time and material. In the sample shown here, you would complete pour-plate counts on the last four dilutions (numbers 3, 4, 5, and 6). Incidentally, one advantage of getting into the habit of using a tenfold serial-dilution pattern is its easy adaptation to a logarithmic scale; each step increases the dilution by one or two logs.

## SOURCES OF ERROR IN SERIAL DILUTIONS

Two common sources of error introduced in the serial-dilution procedure are pipetting errors and errors introduced by failing to mix the diluted samples sufficiently. The pipetting errors tend to resolve themselves with practice. As you gain experience, the reproducibility of your results will improve. If you remember a few rules about pipetting, your errors will be markedly reduced.

1. Always determine the type of pipette you are using and then use it correctly. A *measuring pipette* (or *Mohr*) is designed to accurately deliver the volume indicated between two marks on the pipette. The graduations on a measuring pipette do not extend to the tip. A *serological pipette* is designed to deliver the stated volume by allowing the fluid to flow out of the pipette under the force of gravity. The graduations on this type of pipette always extend to the tip.
2. Always hold the pipette in the vertical position. This way, the meniscus can always be read more accurately.
3. Always use the bottom or top of the meniscus (do not try to estimate the midpoint) to measure the volume. Because of the surface-tension properties of aqueous solutions, the meniscus formed in a pipette is usually convex, and the bottom of the meniscus should be used as your measuring point.

Mixing errors can be avoided by taking care to always mix the samples carefully. You should mix all the tubes the same way—for example, by vortex mixing for one minute, or by shaking twenty times. It really doesn't matter how you mix the sample, as long as you are consistent and careful to ensure complete mixing.

1. Always mix the dilutions carefully.
2. If a vortex mixer is available, vortex each tube for the same amount of time.
3. If you use bottles or screw-cap tubes, shake each bottle or tube the same number of times.
4. The key to accurate dilutions is consistency; mix and pipette each dilution the same way.

The serial-dilution protocols discussed here all assume that 1.0 ml aliquots will be plated out. Sometimes it is convenient to plate out 0.1 ml and make the appropriate mathematical changes in dilution calculation. Using the smaller volume allows you to stretch out your dilution sequence, but it also makes pipetting errors proportionately larger. Most students can pipette 1.0 ml reasonably well, but it takes practice to consistently pipette 0.1 ml accurately.

## POUR-PLATE COUNTING OF BACTERIA AND VIRUSES

If you added 1.0 ml of a suspension of either bacteria or viruses to melted agar (see Chapter 1) and then poured the agar into a petri dish, the bacteria (or viruses) would be immobilized in the agar when it solidified. After a suitable incubation time, the colonies (or plaques) would become visible and you would be able to count them. Figure E24-1 on page 235 shows a typical 1:10 serial dilution/pour-plate-counting protocol. Try to follow the steps as you read the following paragraphs.

Although a single colony may contain more than a million bacterial cells, it can be assumed that all of those cells arose from a single cell. The number of colonies is, therefore, an accurate representation of the number of viable cells (able to grow on the medium used and under the experimental conditions used) present in the sample of fluid that was mixed with the agar.

Since viral growth destroys the host cell and causes it to lyse (dissolve), virus growth will be seen as a clear area in any otherwise solid layer of cells. This cleared area, completely analogous to a bacterial colony, is called a virus plaque. And, like a colony, each plaque is the result of virus replication.

If you did one of these counts on each tube in a serial dilution, you would have completed a plate count or a plaque count. Your results might look like those shown in Table 4-4.

TABLE 4-4. RESULTS OF A PLATE COUNT OF A BACTERIAL OR VIRAL SUSPENSION

| Plate Number | Dilution (Log 10) | Colonies or Plaques |
|---|---|---|
| 1. | −1 | TNTC* |
| 2. | −2 | TNTC |
| 3. | −3 | TNTC |
| 4. | −4 | 321 |
| 5. | −5 | 38 |
| 6. | −6 | 5 |
| 7. | −7 | None |

*Indicates colonies were Too Numerous to Count (the undiluted sample was not plated out).

When counting colonies or plaques, only those plates that have between 30 and 300 colonies are counted. In the example shown in the table, the only "countable" plate is plate #5, the one with the 1:100,000 dilution. It had 38 colonies. Accordingly, the bacterial count of the original sample was 3.8 million bacteria/ml.

(38 colonies × 100,000 dilution

= 3,800,000 bacteria/ml)

The reason for counting plates that have between 30 and 300 colonies is that the most representative samples are obtained when the number of colonies falls within that range. Less than 30 colonies is not considered to be a representative sample (too few organisms). More than 300 colonies is similarly considered nonrepresentative because the overcrowding of the colonies will result in poor growth of the bacteria, and some colonies may not have grown to visible size. Furthermore, technician error increases as the colonies become smaller, more numerous, and more difficult to see.

As mentioned earlier, we assumed that you plated 1.0 ml of each dilution. Had you done otherwise, you would need to make appropriate changes in your calculations. For example, if you plated out only 0.1 ml, and observed the same number of colonies, the number of bacteria in the original sample would be 38 million/ml. Why?

## VARIATIONS IN PLATE-COUNTING PROCEDURES

There are at least two variations on the pour-plate counting procedure that have important applications. These are the surface-plate count and the membrane-filter count. Each of these procedures has important advantages that makes it more suitable for counting under appropriate circumstances.

### Surface-Plate Counting

There are many occasions when it is important that the bacteria not be exposed to any elevated temperature, even that encountered by placing the cells in melted agar. In other instances, when very large numbers of samples must be counted, it may be convenient to prepour the agar into the petri plates and avoid the logistic problems of maintaining very large numbers of melted agar pours.

When surface-plate counting is indicated, the diluted bacterial suspension is spread over the surface of hardened agar rather than being mixed with melted agar and then poured into a petri plate. Typically, 0.1 (one-tenth) ml is placed on the agar and then spread out over the surface with a bent glass rod that has been sterilized by soaking in alcohol and flaming. Since only 0.1 ml of suspension was plated, appropriate changes in calculations must be made. This technique is used in Exercise 16.

### Membrane-Filter Counting

As you have already seen, serial dilutions are useful when the number of bacteria in a sample is very large. Membrane-filter counting techniques make it possible to detect and accurately count small numbers of bacteria in relatively large samples. Also, since the bacteria are not exposed to the heat of melted agar, this procedure is useful for counting heat-sensitive organisms. Membrane-filter counting has proven particularly useful in the bacteriological testing of water (where large volumes can be easily tested for low numbers of coliforms) and in quality-control laboratories (where large volumes can be tested to ensure that they are sterile). Incidentally, as mentioned in Chapter 3, membrane-filtering techniques are often used to sterilize heat-labile fluids.

When a known volume of the sample is passed through a bacteriological-grade filter, any bacteria in the fluid will be retained on the filter. These bacteria will remain viable for a short period of time and will develop into visible colonies if the filter is placed on a pad saturated with nutrient medium. The volume of sample tested is limited only by the capacity of the filtering apparatus.

This technique is very adaptable to specialized counting procedures and, depending upon the type of medium used, can differentially count bacterial populations. For example, membrane-filter counting is routinely used to detect coliforms in water samples containing other bacteria. Membrane-filter techniques are demonstrated in Exercise 24.

## MOST PROBABLE NUMBER ANALYSIS

The most probable number analysis, or MPN, while not commonly used in clinical applications, is extensively used in public health and environmental microbiology.

There are two important advantages of the MPN procedure. First, it is a good screening procedure, useful when a high level of precision is not required. (Do you need to know *exactly* how many bacteria are present, or is it sufficient to know *about* how many are present?) The second advantage is that selective and differential media can be used. This makes it possible to estimate numbers of specific kinds of bacteria, such as coliforms in water. One important limitation of this procedure is its lack of precision. The data obtained from an MPN test is only an estimate of the most *probable* number of bacteria in a given volume of sample.

The MPN test is accomplished by inoculating three sets of tubes with known volumes of sample. Typically, there are three or five tubes of media in each of three sets. The first set usually contains 10.0 ml of double-strength broth; each of the three tubes in the set is inoculated with 10.0 ml of sample. Double-strength broth is used so the medium will not be too dilute when you add an equal volume of sample. The second and third sets of three or five tubes contain 10.0 ml of single-strength broth and are inoculated with 1.0 ml and 0.1 ml of sample, respectively. (See Table 4-5.)

**TABLE 4-5. MOST PROBABLE NUMBER SETS**

| Set | 10 ml of ... | Volume of Sample Used |
|---|---|---|
| First | Double-strength broth | 10.0 |
| Second | Single-strength broth | 1.0 |
| Third | Single-strength broth | 0.1 |

After a suitable incubation period, the pattern of growth obtained is noted and compared to a table of statistically determined most probable numbers. For example, if two of the 10.0 ml tubes *and* one of the 1.0 ml tubes showed growth (as: ++−/+−−/−−−), the most probable number would be 15 bacteria per 100 ml of sample. The MPN procedure estimates the *most probable number* of bacteria likely to be present in 100 ml of the sample. The MPN tables give that value for each of the possible combinations of growth in the three sets of tubes, from all positive (+++/+++/+++) to all negative (−−−/−−−/−−−). Consult the Text and Resource References for additional information about this procedure.

## REFERENCES

### Text References

1. Atlas, Ronald, M., *Microbiology: Fundamentals and Applications.* Chapter 9.
2. Brock, Thomas D., David W. Smith, and Michael T. Madigan, *Biology of Microorganisms,* 4th ed. Chapter 7.
3. Jensen, Marcus, *Introduction to Medical Microbiology.* Chapter 6.
4. Nester, Eugene W., et al., *Microbiology,* 3rd ed. Chapter 4.
5. Stanier, R. Y., E. A. Adelberg, and J. Ingraham, *The Microbial World,* 4th ed. Chapter 5.
6. Stanier, R. Y., et al., *Introduction to the Microbial World.* Chapter 5.
7. Wistreich, George A., and Max D. Lechtman, *Microbiology,* 4th ed. Chapter 6.

### Resource References

1. Finegold, Sidney M., and W. J. Martin, *Bailey and Scott's Diagnostic Microbiology,* 6th ed. Chapter 10.
2. Gerhardt, P., ed., *Manual of Methods for General Bacteriology.* Chapter 11.

# Exercise 1
# Isolation Techniques

If you were asked to write a description of a leaf, you would first sort leaves into groups that had similar characteristics. Then you could begin to describe the shape, the location of the veins, and whatever other characteristics were apparent for each of the groups. You would not collect leaves at random and attempt to write a single description of all the leaves one is likely to find on the forest floor. When you searched for the one type of leaf you wanted, you actually (perhaps subconsciously) used physical appearances to distinguish leaf types—an important and necessary first step in the characterization of any living form.

In microbiology, the procedures that accomplish separation of bacteria into discrete types are referred to as *isolation procedures*. Just as you needed to separate the leaves into discrete groups, you must isolate your bacterial forms into pure cultures. A *pure culture* is a culture consisting of a single type (species or strain) of bacteria, usually derived from a single cell. Unlike the leaf, and unlike most of the life forms we are familiar with, bacteria are too small to see individually without the use of microscopes (Exercises 2, 3, and 4). We must rely on the ability of single cells to grow into a mass of cells large enough to see. Such a mass of cells is referred to as a *colony*.

Suspensions of cells are spread over the surface of a nutrient medium in such a way that each cell will grow into a colony that is physically separated from any other colony. Since colonies are assumed to be clones, isolated colonies represent pure cultures.

In this exercise, we will attempt to isolate mixtures of bacteria into pure cultures. We will also examine the appearance of the colonies and determine the colonial morphology of the organisms studied. It is important to remember that when we isolate a pure culture from a mixed culture, the colonies that grow may contain a bacterium that the microbiologist is seeing for the first time. The recognition of colony types on the basis of their morphology is the critical first step in diagnostic bacteriology; it is the basis of all clinical isolations.

## ASEPTIC TECHNIQUES

Bacteria can be isolated from almost any source; because they exist virtually everywhere, we must be careful not to contaminate our cultures with bacteria from the air or instruments. Obviously, neither do we want to contaminate our work area or anything in it (including ourselves and our laboratory partner) with the bacteria in the culture we are studying.

The precautions used to reduce the chances of contamination are collectively referred to as *aseptic techniques*. If these precautions are properly used, the chance of contamination can be reduced to almost zero.

Your laboratory instructor will demonstrate these precautions as they are encountered. They are important and essential for the safety of yourself and your laboratory partners. Of course, they are also important to prevent you from identifying a contaminant rather than an unknown or to keep you from wasting your time identifying a contaminant instead of the significant organism from a clinical sample.

These precautions may include some or all of the following:

- Flaming of inoculating loops and needles and proper handling of sterile pipettes;
- Prevention of splattering caused by using needles and loops that have not cooled after flaming. The most common source of laboratory contamination and infections is the aerosols created by splattering;
- Flaming of the mouths of test tubes and other culture vessels and maintaining sterility of glassware;
- Proper disposal of contaminated instruments, pipettes, and cultures;
- Proper cleaning of work area before and after use;
- Safety precautions in cases of accidental spills;
- Proper methods for washing hands and the use of germicidal soaps;
- Use of absorbent paper mats (may be wet with disinfectant) at work areas to absorb spills.

## ISOLATION PROCEDURES

There are two isolation procedures commonly used in microbiology. One, the *streak plate,* is probably the single most-used procedure in any microbiology laboratory. The mixture of bacteria is spread over the surface of a solid nutrient medium so that isolated colonies develop. The second procedure, the *pour plate,* is less commonly used for routine isolation, but is frequently used to estimate the numbers of bacteria in a sample. It requires that a portion of the sample be mixed with the melted medium, which is then allowed to solidify. If the sample was properly diluted, isolated colonies will develop after incubation.

## STREAK-PLATE PROCEDURES

The successful streak plate will have isolated colonies that can be transferred to fresh media with confidence that the cultures are pure.

1. A small amount of colony material is picked from the source culture with an inoculating needle. If the source culture is liquid, a needle or loop is used and simply dipped into the sample (using aseptic techniques, of course).

2. The needle or loop is then streaked over the surface of a nutrient medium in such a way as to spread the sample over the medium. As the needle or loop moves across the surface of the medium, most of the bacteria are deposited at the beginning of the streak. Toward the end, however, the remaining bacteria are placed on the medium so that when they grow, isolated colonies will develop.

3. Failure to obtain isolated colonies can almost always be traced to one of two common mistakes:

   a. Using *too much inoculum*: If you transfer too much sample to the streak plate, there will not be sufficient surface area to spread out the bacteria. A good rule of thumb is that if you can see more than just a speck of culture material on the end of the needle, you have too much. Do not use a loop for transfer from agar and do not use a loop if the liquid sample is very turbid.

   b. Using *too little surface area*: A single wavy line down the center of the plate will not produce isolated colonies. *Use as much of the surface of the medium as possible.* Place the streaks as close together as possible without overlapping them. The pattern you choose is not nearly as important as whether or not you get isolated colonies.

"Four quadrant" streak plate

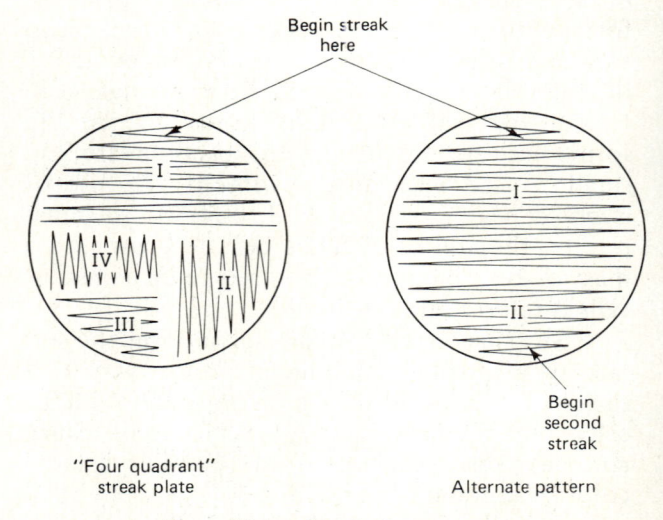

Alternate pattern

**Figure E1-1** Streak-plate patterns.

4. The needle and loop will dig into the medium if too much pressure is applied when making the streak. There should be only enough pressure to "steer" the instrument; rely on its weight and balance to provide most of the downward pressure. Practice on the countertop (you do not need to flame the needle this time). If you do cut into the medium (and only practice will make it unlikely that you will), stop the streak and carefully remove the needle or loop to minimize splattering when the tension on the instrument is released.

## POUR-PLATE PROCEDURES

The pour plate differs from the streak plate in that the sample is mixed with melted medium, which is then poured into a petri dish. After the medium solidifies, the bacteria are held in place by the solidified medium; isolated colonies develop after appropriate incubation.

1. An aliquot of a liquid sample is transferred with a sterile pipette to liquid medium that is held in a water bath at about 50°C. The medium and sample are then gently mixed (gently enough so that *no bubbles* are formed) and poured into a petri dish. After the medium solidifies, it can be incubated and the isolated colonies observed and selected.

2. One of the most common problems encountered by beginning microbiologists is that of determining the correct dilution needed to produce isolated colonies. Experience is the best teacher, but a good rule is to try to use dilutions that will bracket the correct one. For example, most experienced microbiologists will use three or four dilutions just to be sure of getting plates that will have between 30 and 300 colonies.

3. The simplest way of diluting your sample is to transfer a small amount (typically 1.0 ml) of the sample into a known volume of either saline (at 0.9%) or water. If 9.0 ml of diluent is used, you will have made a 1:10 dilution, and if you use 99 ml, a 1:100 dilution will result. These dilution steps can be combined in any order to dilute the sample as much as needed. In this exercise, we will make several 1:10 dilutions of the sample and transfer one loopful of the diluted samples into the medium.

4. If the pour plates are to be used for counting of bacteria, they should have between 30 and 300 colonies on them for the counts to be considered valid. More than 300 colonies usually means an overcrowded plate where some colonies are too small to see or did not grow at all. What is the reason for the lower limit of 30?

## QUANTITATIVE ESTIMATIONS

If you know the volume of the original sample used to make the first dilution, and the final dilution used, you can easily determine the number of bacteria present in the sample. You simply have to multiply the three variables together to obtain the answer. Most microbiologists will report the result as the number of "bacteria/ml" or "colony-forming units (CFU)/ml." The use of quantitative pour plates will be more extensively covered in Exercise 24.

The streak plate can be used to obtain a very rough approximation of the number of bacteria in the original sample. If the plate and streak pattern is mentally divided into four quadrants, the appearance of colonies in each quadrant can be used as a relative measure of the numbers present in the original sample. The most common system used is a 4+, 3+, 2+, and 1+ system, indicating in which quadrants the colonies were observed. For example, a 3+ would indicate that colonies appeared in the first three quadrants, but not in the fourth. This type of estimation is almost universally used by clinical laboratories. For example, the laboratory report will read "3+ alpha-hemolytic *Streptococcus*" or "1+ beta-hemolytic *Streptococcus*," and so on.

## COLONIAL MORPHOLOGY

Bacterial colonial morphology refers to the physical appearance of isolated colonies. Remember, when a colony is grown on an isolation plate, you will often be observing any physical characteristics of the bacterial isolate for the first time. These physical characteristics are often specific for the type of bacterium making the colony and can be used as a means of recognition. Colony morphology is, however, influenced by the medium and other growth conditions. The colonial morphology of the same bacterium may vary on different media or under differing environmental conditions.

**Topography or elevation of colonies**

Convex · Umbonate ("fried egg") · Plateau · Flat
Raised · Irregular · Cratered ("dimpled")
Raised, spreading edge · Flat, with raised margin · "Gumdrop"-like · Growth into the agar

**Edge morphology of colonies**

Smooth entire · Rhizoid "woolly" · Raised, spreading edge · Lobate
Irregular wavy · Filamentous branched · Fibrous "woolly"

**Complex colonial morphologies**

Irregular and spreading · Round, scalloped · Wrinkled · Filamentous rhizoid

**Patterns of growth in broth**

Surface-diffuse · Settled-diffuse · Diffuse · Pellicle

Some Cultural Characteristics of Bacteria

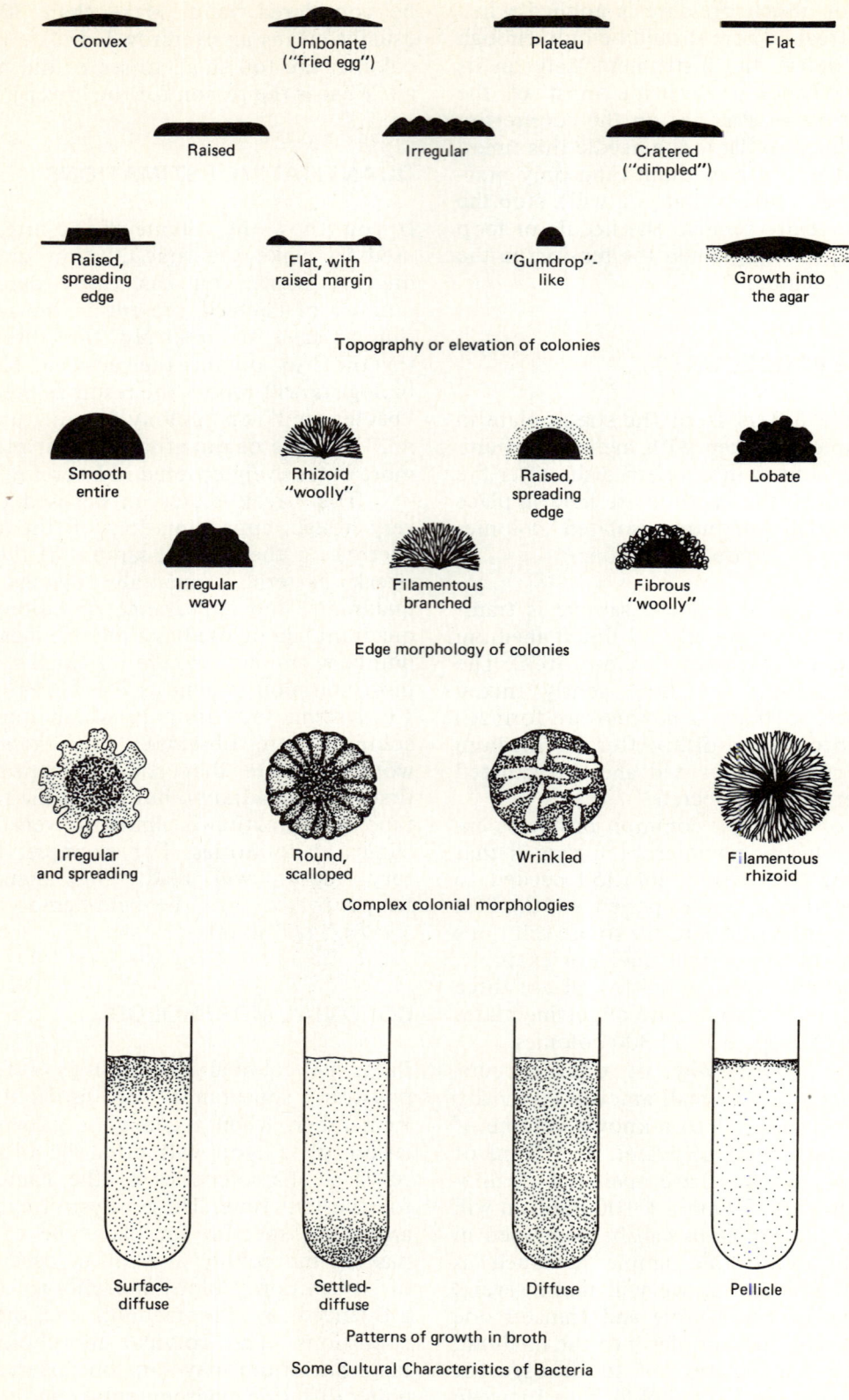

Figure E1-2

*Note*: Colonial morphology is usually determined on colonies growing on the surface only. Colonies embedded in the medium, as with a pour plate, are surrounded by the medium and are limited by the physical characteristics of the medium. Colonies submerged in medium tend to be lenticular, or football-shaped, because the colony growth tends to split the medium, allowing growth only in the "bulge" thereby produced.

**How to Recognize Colonies**

Colony recognition is an individual accomplishment. We can show you where to look for differences, but *you* must recognize those differences. For example, if everyone in your class was asked to write a description of your laboratory instructor, would you all come up with identical descriptions? Even though your descriptions are different, were any of them wrong? Did any of them fail to accomplish the objective of picking out the laboratory instructor from a group? We will suggest a list of characteristics that you can use to identify colony types. Whether you use them all or not is up to you — so long as you are able to recognize that colonies do differ and then are able to pick them out of a mixed group of colonies.

Any characteristic that allows you to differentiate and recognize colony types may be used. The list below is meant to be a guide — use whichever characteristics are necessary for *you*, not your partner, to recognize the colony types.

**Colony Physical Characteristics (Fig. E1-2)**

1. Colony shape and topography:

   | Convex | Concave |
   |---|---|
   | Flat | Dimpled |
   | Center plateau | "Fried-egg" shape |

2. Colony consistency:

   | Smooth | Mucoid |
   |---|---|
   | Dry | Rough |
   | Fibrous | Granular |

3. Colony edge:

   | Spreading | Smooth |
   |---|---|
   | Rhizoid | Lobate |

4. Colony color:

   Color of the colony
   Color of the medium around the colony
   Iridescence

5. Changes in medium surface:

   Some bacteria dissolve agar and will appear to sink into it.

## LABORATORY OBJECTIVES

In this exercise you will learn how to isolate mixtures of bacteria into pure cultures and to recognize the specific colony morphology of some bacteria. You should:

1. Understand the principles of bacterial isolation.
2. Be able to explain the streak plate and the pour plate and the differences between the two.
3. Be able to explain how each can be used quantitatively.
4. Be able to write a description of some bacterial colonies.
5. Be able to identify colonial types in a mixture.

## EXPERIMENTAL DESIGN OF THIS EXERCISE

You will be required to make at least seven streak plates and at least one set of pour plates. The streak plates will be made from pure cultures (at least five) and from a mixture or mixtures containing those same five bacteria. After incubation, you will determine the colonial morphology of the pure cultures and then attempt to find them in the mixed culture.

The pour plates will be made from either the pure cultures (if they are in broth) or from the mixture. You should compare the colonies produced by submerged bacteria with those growing on the surface.

## MATERIALS NEEDED FOR THIS LABORATORY

1. At least seven nutrient agar plates. The plates should have been prepared 24 hours ahead of time and incubated. This will

allow the plates to dry so that there is no surface moisture to ruin your streak plates (especially with motile bacteria). It is also a good quality-control practice to eliminate plates that were accidentally contaminated when they were prepared.

2. Twenty-four-hour agar slant cultures of:
   *Staphylococcus epidermidis*
   *Escherichia coli*
   *Bacillus subtilis*
   *Pseudomonas aeruginosa*
   *Serratia marcescens* (pigmented strain)

3. Mixed cultures of the above bacteria. These should be prepared immediately before the laboratory in either saline or nutrient broth. The suspensions should be dense enough to ensure growth of all colony types. The recommended mixtures include:
   Mixture One:
     *Staphylococcus epidermidis*
     *Escherichia coli*
     *Bacillus subtilis*
   Mixture Two:
     *Serratia marcescens*
     *Pseudomonas aeruginosa*
     *Staphylococcus epidermidis*

4. Five nutrient agar deeps. These will be used for the pour plates and will need to be melted and then held at 50°C to keep them liquid.

5. Five sterile petri dishes.

6. Six water-dilution blanks, containing 9.0 ml of sterile water.

7. Sterile pipettes.

## LABORATORY PROCEDURE

1. Obtain and label the petri dishes you will need for the streak plates. You will need one for each pure culture and one for each mixed culture. At least seven will be required.

2. Using aseptic techniques as demonstrated by the instructor, make streak plates of all the cultures. Remember that you should not use too large an inoculum and that you should use as much of the surface of the medium as possible. Avoid digging into the medium. If you make a mistake, repeat the streak plate with a fresh plate.

3. Obtain and label the dilution blanks (six) and the sterile petri dishes (five). Obtain the necessary sterile pipettes. *Do not open the dishes.*

4. Obtain five tubes of melted nutrient agar deeps *when you are ready to complete the pour plates.* It may be necessary to set up a water bath to keep the tubes liquefied. Remember that agar solidifes at about 48°C, but must be heated to about 100°C to melt. If your medium solidifies in the tubes, the experiment is ruined and you must begin again.

   You should have everything set up before you obtain the agar deeps. Once they are removed from the water bath, they will solidify rapidly. *Examine the table and flow diagram (Fig. E1-3) as you read these instructions.*

5. Using aseptic techniques, transfer 1.0 ml of your sample to the first dilution blank; mix well. (You now have a 1:10 dilution.)

6. Transfer 1.0 ml from the first dilution blank into the second blank, mix it well by shaking, and then transfer 1 ml to each subsequent dilution blank until all of the dilution blanks have been inoculated.

7. From this point on, you will be making pour plates of each dilution. *Now is the time to ensure that the medium is ready and petri plates are in place.*

8. Transfer one loopful of bacterial suspension from dilution blank #2 to agar tube #1. Gently mix the medium and immediately pour it into a petri dish. Allow the petri dish to remain undisturbed while you continue.

9. Repeat this step until all the tubes of soft medium and petri dishes have been used. You should have five plates by the time you are finished.

| Dilution Blank No. | Agar Deep No. | Petri Plate No. |
|---|---|---|
| 1 | Not plated | Not plated |
| 2 | 1 | 1 |
| 3 | 2 | 2 |
| 4 | 3 | 3 |
| 5 | 4 | 4 |
| 6 | 5 | 5 |

10. After the medium has solidified, place all plates in the incubator. Allow to incubate for 24 hours.

Transfer 1.0 ml from one tube to another

Dilution sequence → Sample | No. 1 1:10 | No. 2 1:100 | No. 3 1:1,000 | No. 4 1:10T | No. 5 1:100T | No. 6 1:1M

Pour plate protocol → Dilution blank → Transfer 1.0 ml → Agar deep → Pour into plate → Petri plate

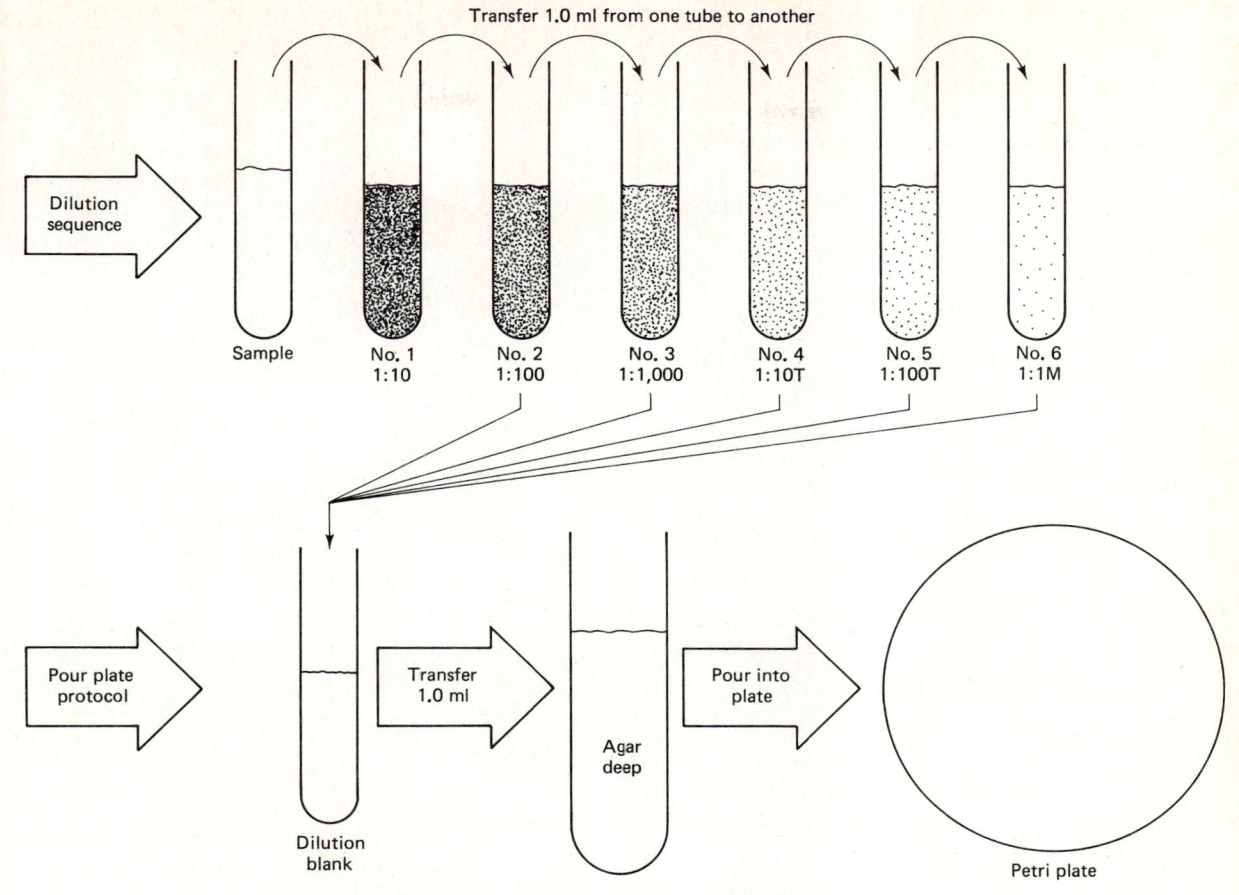

**Figure E1-3** Dilution and plating protocol for pour plates.

11. Record your observations. Write a description of each colony type and try to identify each type in the mixtures.
12. If you did not get isolated colonies, try to determine why. Repeat the experiment until you have obtained isolated colonies.
13. Examine the pour plates. Compare the colonies with those obtained on the streak plates.
14. Record your results in the Laboratory Report Form. Be sure to fill in the Bacterial Characteristics Form (Appendix 6).

## REFERENCES

### Text References

1. Atlas, Ronald M., *Microbiology: Fundamentals and Applications.* Chapter 2.
2. Brock, Thomas D., David W. Smith, and Michael T. Madigan, *Biology of Microorganisms,* 4th ed. Chapters 7 and 18.
3. Jensen, Marcus, *Introduction to Medical Microbiology.* Chapter 6.
4. Nester, Eugene W., et al., *Microbiology,* 3rd ed. Chapter 4.
5. Stanier, R. Y., E. A. Adelberg, and J. Ingraham, *The Microbial World,* 4th ed. Chapters 1 and 2.
6. Stanier, R. Y., et al., *Introduction to the Microbial World.* Chapters 1 and 2.
7. Wistreich, George A., and Max D. Lechtman, *Microbiology,* 4th ed. Chapter 6.

### Resource References

1. Finegold, Sidney M., and W. J. Martin, *Bailey and Scott's Diagnostic Microbiology,* 6th ed. Chapter 4.
2. Gerhardt, P., ed., *Manual of Methods for General Bacteriology.* Chapters 8, 9, and 20.
3. Lennette, E. H., ed., *Manual of Clinical Microbiology,* 3rd ed. Chapter 6.
4. McGonagle, L. A., *Procedures for Diagnostic Microbiology.* Pages 137–138.

# LABORATORY PROTOCOL

## Exercise 1—Isolation Techniques and Colonial Morphology of Bacteria

Check each step when you complete it.

### First Day

_____ 1. Obtain and label the necessary number of nutrient agar plates to complete the streak plates. You will need at least seven plates.

_____ 2. Make one streak plate of each of the pure cultures. Be sure to use a small amount of inoculum and to use as much of the surface of the medium as possible.

_____ 3. Make a streak plate of each of the mixed cultures. You will be trying to isolate the mixture into pure cultures.

_____ 4. Obtain six dilution blanks; label them 1 through 6. Obtain five sterile petri dishes; label them 1 through 5 (see table on page 30).

_____ 5. Make the requisite dilutions of the sample. Do this by transferring 1 ml of sample into blank #1 and mix well. Then transfer 1 ml from the first blank into each subsequent blank (#2 through #6), mixing well between each transfer, until all of the six dilution blanks have been seeded with bacteria.

_____ 6. Obtain the melted nutrient agar deeps. Be certain that everything is ready to complete the pour plates and that the medium cannot solidify while you are making the dilutions (do you need a water bath?).

_____ 7. Transfer one loopful from blank #2 to the first tube of melted medium. Mix the medium gently and pour into the petri dish labeled #1.

_____ 8. Repeat step 7 until you have completed all the pour plates. Transfer one loopful from blank #3 to tube #2 and then pour the seeded medium into the appropriate plate, and so on until you have finished the series.

_____ 9. Incubate all plates for 24 hours.

_____ 10. Write a description of the colonial morphology for each of the pure cultures.

_____ 11. Identify the individual colony types in each of the mixed cultures. Write a two- or three-sentence summary of your findings.

_____ 12. Examine the pour plates. Compare the colonies that developed on the pour plates with those observed on the streak plates.

_____ 13. Compare the colonies that are submerged in the medium with those that grew on the surface.

_____ 14. Complete the Laboratory Report Form, being sure to fill in the Bacterial Characterization Form (Appendix 6).

# NOTES

Name: _____

## LABORATORY REPORT FORM

### Exercise 1—Isolation Techniques and Colonial Morphology of Bacteria

1. Enter the colonial morphology on one Bacterial Characteristics Form for each organism studied in this Exercise. These forms are provided in Appendix 6. Remember that the data on the forms will be accumulated over several exercises and added to the forms as it is obtained.
2. Write a short description of each of the colony types observed in this exercise. (Use drawings if you would find them helpful.)

*Staphylococcus epidermidis*:




*Escherichia coli*:




*Bacillus subtilis*:




*Pseudomonas aeruginosa*:

35

*Serratia marcescens*:

**ANSWER THE FOLLOWING QUESTIONS**

1. What is an *aliquot* of a sample?

2. How could the pour-plate procedure be used for counting the number of bacteria in a sample?

3. Why do the submerged colonies look different than the one growing on the surface of the medium?

4. Explain a laboratory report of throat culture that reports "2+ beta hemolytic *Streptococcus*."

5. How would you know if a bacterial culture excreted a water-soluble pigment?

6. How would you know if a bacterial culture produced a pigment, but did not excrete it?

7. Give two reasons why you should preincubate nutrient agar plates.

# Exercise 2
# Preparation of Smears, Simple Stains, and Wet Mounts

In the previous exercise we addressed the recognition of bacterial colonies. Colonies are, however, populations of individual cells that happen to be grouped into a mass that is large enough to see. A colony is the visible evidence of the presence of a very large number of individuals, and although the morphology of that colony may be distinctive, we still need information about the individual members of that population. We must be able to examine single cells; to do this, we have to stain them and view them through a microscope.

## BACTERIAL CELL MORPHOLOGY: FOUR BASIC CELL SHAPES

- Cocci: Cocci are spherical cells. Some cocci, especially those that typically form pairs, may be somewhat flattened on the adjacent sides.
- Rods: Rods, or bacilli, are cylindrically shaped, straight cells that appear rigid. Typically, but not always, the diameter of the cell is constant throughout its length (variations in cell morphology will be discussed in a later section).
- Vibrios: Vibrios are curved rods whose curvature is always less than a half-circle (less than 180 degrees). As with the bacilli, the diameter is generally constant over the length of the cell.
- Spirals: Curved cells whose curvature exceeds that of a half-circle are characterized as spirals rather than vibrios. These cells usually have several to many complete turns, often resembling small springs when viewed from the side.

## VARIATIONS IN CELL MORPHOLOGY AND CHARACTERISTIC CELL GROUPINGS

Two additional morphological characteristics have proven to be helpful in clinical applications. These characteristics include variations in the cell shape and variations in the clustering of cells after they divide. Although most bacteria separate from each other after division is complete, some tend to remain together. Often, this tendency to remain together produces cell groups characteristic of the species or genus.

### The Genus *Corynebacterium*: The Diphtheroids

The corynebacteria, or diphtheroids, are a prominent component of the normal flora of the skin and upper respiratory tract. Some members of the genus are pathogenic. Diphtheroids have a typical and distinctive morphology because of their tendency to remain attached to each other after division. They do not, however, simply tend to form longer and longer filaments, but rather snap together, producing clusters of cells that have been likened to "palisades" and/or "Chinese characters." In addition, the cells are often thicker at one end and appear to be club-shaped. The combination of palisade or Chinese-character clusters and club-shaped cells is quite distinctive. An experienced microbiologist is able to recognize diphtheroids almost immediately.

## The Cocci:
### Characteristic Cell Groupings

Cocci, like all bacteria, divide by transverse binary fission. Unlike other bacteria, however, some cocci can vary the plane of division and produce clusters of cells that may take the form of long strands, pairs, packets of four (tetrads) or eight, or random clusters. Since these groupings are characteristic for the genus, they do have diagnostic value and are immediately recognized by the experienced microbiologist.

Streptococci characteristically form either filaments several cells long or pairs of cells frequently referred to as *diplococci*. Each plane of division is parallel to the previous one.

*Neisseria* and *Branhamella* are genera of cocci that form pairs of cells (diplococci). These two genera also have typically flattened adjacent sides, so much so that they resemble kidney beans arranged with the flattened sides adjacent to each other.

Staphylococci and micrococci typically form random clusters of cells that have been likened to grapelike clusters. This distinction between the streptococci and the staphylococci is frequently used as an important part of the diagnostic protocol for the cocci. The plane of division in these cells appears to be randomly selected.

Some of the micrococci divide in alternating planes and produce packets of four or eight cells. These groupings of cells are formed when the plane of each division is at right angles to the previous one.

The examples given here are, of course, not exhaustive. They do represent examples that have significance in clinical situations.

## STAINING PROCEDURES

A chemical to be used as a stain for biological material must have at least two properties: It should be intensely chromatogenic (intensely colored) and it must react with some cellular component. Although these requirements seem simple and obvious, the development of histological techniques did require substantial effort, and the statement that they must react with some cellular component is deceptively complex and somewhat misleading in its simplicity. Nevertheless, all biological stains have these properties and, when used correctly, enable us to visually examine cells with a considerable level of resolution and definition.

Simple stains are ones that can be used as general, all-purpose stains. As a rule, they stain biological material indiscriminately, producing a uniformly stained cell. It is usually necessary to expose the cells to the stain for a short period of time only and then to remove the excess stain by rinsing before we examine the cells with a microscope. In the next exercise we will see how selective reactions can be used to differentially stain certain types of cells and/or parts of cells.

## PREPARATION AND STAINING OF SMEARS

The preparation of a microscope slide with stained biological matter on it is a two-step process. These steps are:

1. Preparation of smears: A smear, in the context used here, is a slide that has had bacteria placed on it and which has been treated to cause the bacterial cells to adhere to the slide.
2. Staining of smears: Once the bacteria are on the slide, and have been made to adhere to it, they must be stained. This involves exposing the cells to the stain, allowing them to react with it for the required period of time, washing the excess stain from the slide, and then drying it.

## UNSTAINED PREPARATIONS: WET MOUNTS AND HANGING DROPS

It is often necessary to examine cells while they are still alive. For example, one method for the determination of bacterial motility requires that we watch the bacteria "swim." There are two procedures that allow this, one being a variation of the other.

1. Wet mounts: A wet mount is prepared by simply placing a loopful of broth on a coverslip and then inverting it on a slide so that the drop of broth is between the slide and the coverslip. If the preparation is to be examined over a long period of time, petroleum jelly may be used to seal the edges and prevent drying.
2. Hanging drops: If the coverslip, prepared as above, is carefully placed on a slide that has a depression molded into it, the drop of broth will hang in the space of the depression, thus the name *hanging drop*. This technique is useful when relatively thick cells, such as protozoans, must be examined.

## LABORATORY OBJECTIVES

The visual examination of bacterial cells is an important part of their characterization. It is important that you know how to do this and that you are reasonably comfortable with high-magnification compound microscopes. In this exercise, you will:

1. Review the use of oil-immersion optics and the general use and care of microscopes.
2. Prepare bacterial smears and wet mounts for microscopic study.
3. Stain bacterial smears and observe cell morphology and typical groupings.
4. Understand the basic principles underlying staining procedures.

## MATERIALS NEEDED FOR THIS LABORATORY

1. Twenty-four- or 72-hour cultures of:
    *Staphylococcus epidermidis,* slant and broth
    *Escherichia coli,* slant and broth
    *Bacillus megaterium,* 72-hour slant
    *Bacillus subtilis,* 72-hour slant
    *Corynebacterium xerose,* slant
2. In addition to, or as alternatives to the above cultures, prepared slides may be used. It is often helpful for students to use prepared slides as they familiarize themselves with the microscope. Prepared slides also offer convenient opportunities for the examination of some fastidious or pathogenic forms.
3. Staining reagents:
    Crystal violet
    Methylene blue
    Safranin
4. Miscellaneous reagents and materials:
    Petroleum jelly
    Glass slides (frosted ends preferred). The slides must be carefully cleaned and free from residual oils.
    Coverslips

## LABORATORY PROCEDURE

Aseptic techniques must be used throughout this exercise. This is especially important with the wet mounts because the bacteria are still alive. *Follow all instructions carefully and pay special attention to your instructor when these techniques are demonstrated for you.*

The stains used in this exercise will also stain your fingers and clothing. Some jewelry may also be discolored. *Use appropriate caution.*

1. Prepare smears of all the cultures used in this exercise. You should prepare three sets of smears, one for each of the stains used.
    a. The glass slides must be especially clean. If there is any residual grease on the slide, the smears will be irregularly spread over the glass and it will be difficult to observe bacterial morphology.
    b. If you are using slides with a frosted end, you can use a pencil to label the slide; the frosted end will serve as a marker to keep the slide properly oriented (right side up and right to left).
2. Preparation of smears from agar cultures:
    a. Place a small drop of water on the slide where you want the smear to be located. Your inoculating loop is a handy means of applying the proper amount of water.
    b. Using your inoculating needle (and aseptic technique), transfer a small amount of culture to the drop. Emul-

sify the culture in the drop, producing a moderately turbid, but not opaque, emulsion. Spread it evenly over an area about one-half inch in diameter. Be sure to flame your needle before *and* after use.
   c. Repeat as needed until three or four smears have been prepared on the slide.
   d. Allow the slide to air dry. *Do not heat the slide until it is completely dry.*
   e. Heat fix the bacteria to the slide by passing the slide through the flame of a Bunsen burner several times. The slide should be moderately uncomfortable when touched to the back of the wrist. If you do not heat the slide enough, the bacterial cells will be washed off during the staining process.
3. Preparation of smears from broth cultures:
   a. Use your inoculating loop to transfer a loopful of a broth culture to the microscope slide. Use aseptic technique and remember to flame the loop before and after use.
   b. Spread the drop over an area about one-half inch in diameter.
   c. If the broth culture is not very turbid, you may need to apply two or three droplets, one over the other.
   d. Repeat as needed, placing three or four smears on each slide.
   e. Heat fix, as described above.
4. After the slides have cooled, place them on a staining rack located over either the sink or over a staining dish or tray.
5. Flood the slides with stain. Make sure each smear is covered completely. Do not allow the smears to dry; add additional stain, as needed. Allow the smears to stain according to the times shown:

    Crystal violet:    30 seconds
    Methylene blue:  3-5 minutes
    Safranin:         3-5 minutes

6. Rinse the excess stain from the slide by holding it under a stream of gently flowing cold water. Do not allow the stream of water to fall directly on the smear. Rather, allow water to flow over the smear.
7. Remove excess water from the slide by tapping a long edge of the slide on some paper toweling.
8. Allow the slide to air dry. Some workers prefer to blot the slide with paper toweling. If your instructor allows this, place the slide on a sheet of toweling, stained side up, and fold the toweling over the slide. Press down until the toweling absorbs the water. Do not rub the slide with the paper toweling.

*Note*: Staining procedures, especially simple stains, do not always kill the bacteria (e.g., spore-formers), and blotting the slide may transfer viable bacteria to the toweling. Air drying is the procedure of choice. If you do blot the slide, the paper toweling should be discarded in a careful manner.

9. Examine all the smears with the oil-immersion objective. Write a description of each cell type and complete a Bacterial Characteristics Form for each organism.
10. Prepare a wet mount of each broth culture. Carefully transfer a loopful of culture to the center of a coverslip. Using forceps, invert the coverslip over the center of a clean glass slide.
11. Lower the coverslip onto the slide in such a way that one edge touches first. The cover slip can then be lowered onto the slide. This procedure will minimize the formation and retention of bubbles in the broth droplet.
12. If you intend to examine the wet mount for more than several minutes, you should seal the edges with petroleum jelly. Using a toothpick, carefully place a small bead of jelly (about 1.0 mm) around the four sides of the coverslip *before* placing the loopful of broth on the slide.
13. Invert the coverslip and position it as described above. The jelly should form a seal when you lower the coverslip onto the slide. A small amount of pressure may be needed to completely seal the edges.
14. Examine the wet mount with the high-power and oil-immersion objectives of the microscope. Determine if the bacteria are motile by observing directional movement (as opposed to simple Brownian movement). *Remember that the bacteria are still*

*viable. Dispose of the wet mount in a safe manner.*

15. Complete the Laboratory Report Form for this exercise.

## REFERENCES

### Text References

1. Atlas, Ronald M., *Microbiology: Fundamentals and Applications*. Chapter 2.
2. Brock, Thomas D., David W. Smith, and Michael T. Madigan, *Biology of Microorganisms,* 4th ed. Chapter 2. Appendix 5.
3. Nester, Eugene W., et al., *Microbiology,* 3rd ed. Chapter 3.
4. Wistreich, George A., and Max D. Lechtman, *Microbiology,* 4th ed. Chapter 4.

### Resource References

1. Finegold, Sidney M., and W. J. Martin, *Bailey and Scott's Diagnostic Microbiology,* 6th ed. Chapters 3 and 43.
2. Gerhardt, P., ed., *Manual of Methods for General Bacteriology*. Chapter 3.
3. Lennette, E. H., ed., *Manual of Clinical Microbiology,* 3rd ed. Chapter 98.
4. McGonagle, L. A., *Procedures for Diagnostic Microbiology*. Pages 76–77, 119.

## LABORATORY PROTOCOL

### Exercise 2—Preparation of Smears, Simple Stains, and Wet Mounts

Check each step when you complete it.

### First Day

_____ 1. Prepare the smears as directed in the exercise. Pay proper attention to aseptic techniques. You should complete three sets of smears, one for each of the three stains.

_____ 2. Remember to heat fix and air dry, as appropriate.

_____ 3. Stain each smear, leaving the stain on the slide for the times shown (some experimentation with staining times may be needed):

>  Crystal violet:    30 seconds
>  Methylene blue:    3-5 minutes
>  Safranin:          3-5 minutes

_____ 4. Rinse and air dry as directed in the exercise.

_____ 5. Complete a microscopic study of each culture and record the results on the Bacterial Characteristics Form, using a separate form for each organism.

_____ 6. Prepare a wet mount of each broth culture. Take appropriate precautions. *Remember—the bacteria are still viable.*

_____ 7. Examine the wet mount with the high-power and oil-immersion objectives. Determine if the cultures are motile and record the results on the form provided.

_____ 8. Complete the Laboratory Report Form for this exercise.

# NOTES

Name: _____

## LABORATORY REPORT FORM

### Exercise 2—Preparation of Smears, Simple Stains, and Wet Mounts

1. Complete one Bacterial Characteristics Form for each organism. These forms are provided in Appendix 6.

2. Write a short description of each of the staining procedures used in this exercise.

**ANSWER THE FOLLOWING QUESTIONS**

1. Why is it necessary to heat fix bacterial smears?

2. What is Brownian movement and how is it distinguished from true motility?

3. Why should the slides be carefully cleaned before attempting to make a smear?

4. White a description of a *diphtheroid*; a *diplococcus*; a *vibrio*; and a *bacillus*.

5. What is the organ of motility of the bacteria studied in this exercise?

# Exercise 3
# Differential Staining Procedures

Simple stains, as shown in the last exercise, stain biological materials indiscriminately. Fortunately, by modifying the staining procedure or by using either special stains or added chemicals, it is possible to differentially stain bacteria. In some cases, only certain *types* of cells will be stained, while in other cases, only certain *parts* of cells will be stained. These procedures, which allow us to stain different cells or components different colors, are referred to as *differential staining* procedures.

Differential staining, unlike simple staining, requires at least three component factors or steps to make the procedure differential. These include:

- Primary stain: The primary stain is the stain that is used to color the "target" cells or cell parts. It is the staining reagent that actually stains the cell or cell component that you want to examine.
- Mordant and/or selective treatment: A mordant is a chemical that reacts with the primary stain and with the cell or component you want to see. Its purpose is to enhance the retention of the primary stain. A selective treatment, on the other hand, is an added step or treatment that takes advantage of some special cell characteristic and results in retention of the primary stain by the target cells or cell components. In some procedures both a mordant and a special selective treatment are used.
- Counterstain: The counterstain is usually a simple stain that is used to stain everything that was not stained by the primary stain. It is generally a contrasting color.

You will use three differential stains in this exercise—the *gram stain,* the *acid-fast stain,* and the *metachromatic-granule stain.* These staining techniques are among the most-often-used staining procedures in the clinical microbiology laboratory. They will serve as examples of differential staining procedures, but do not represent an all-inclusive list of the available staining techniques. The next exercise describes other staining procedures that are less commonly used.

## THE GRAM STAIN

The gram stain allows rapid distinction between two major groups of bacteria on the basis of the structure and composition of their cell surface membranes and cell walls. It is probably the single most commonly used staining procedure in microbiology. When microbiologists are discussing a bacterial species, the gram reaction is almost always given. For example, you refer to a "gram-positive coccus," or a "gram-negative rod." Virtually all diagnostic protocols used in microbiology begin with the determination of the shape and gram reaction.

| Primary Stain: | Crystal violet |
| --- | --- |
| Mordant: | Iodine |
| Selective treatment: | Ethanol/acetone rinse |
| Counterstain: | Safranin |

The crystal violet is applied to a smear and allowed to stain the cells for about 30 seconds. The crystal violet is then rinsed off the slide with the iodine solution and the smear flooded with additional mordant. After about two minutes, the smear is rinsed with either ethanol or a mixture of ethanol and acetone. The solvent will remove the crystal violet–iodine complex from gram-negative cells only. The counterstain, safranin, is then used to stain all cells that have not retained the crystal violet.

> Gram-positive cells: Stain purple or dark blue
> Gram-negative cells: Stain pink or red

## THE ACID-FAST STAIN (ZIEHL-NEELSEN)

The genus *Mycobacterium* is one of a very few genera that stain positively with the acid-fast stain. These bacteria have a waxy component in their cell wall that makes the wall almost impermeable to solutes dissolved in water. The wall can be made permeable by heating. The genus *Mycobacterium* includes two important pathogens: *M. tuberculosis* and *M. leprae*. The presence of acid-fast bacteria in sputum or tissue is considered presumptive evidence of either tuberculosis or leprosy. Thus, the acid-fast stain obviously has great significance in medical microbiology.

> Primary stain: Carbol fuchsin
> Selective treatment: Heating cells and acid-alcohol rinse
> Counterstain: Methylene blue

The acid-fast staining procedure that will be used in this exercise is referred to as the Ziehl-Neelsen procedure. There is also a fluorescent staining technique being used increasingly in diagnostic laboratories. Fluorescence microscopy is described in Chapter 1. In the Ziehl-Neelsen procedure, a smear is flooded with the carbol fuchsin stain and heated with a Bunsen flame or a steam bath for about five minutes. After the slide cools, it is rinsed with acid-alcohol until the dye no longer flows from the smear. The slide is then gently rinsed with water, and then stained with methylene blue. The acid-alcohol will remove all carbol fuchsin that has not been "trapped" inside acid-fast cells.

> Acid-fast bacteria: Stain pink
> Non-acid-fast cells: Stain blue

## THE METACHROMATIC-GRANULE STAIN

Metachromatic granules, or *volutin,* are storage crystals produced by some bacteria when there is excess phosphate and carbohydrate (a special medium is usually needed) in their growth medium. The storage crystals are deposits of cellular phosphate and are large enough to see when stained. Although several kinds of bacteria produce these granules, their production is not common, and those bacteria that do so are relatively distinctive. One of the most commonly encountered bacteria that do produce the granules is *Corynebacterium diphtheriae*. The detection of diphtheroids with metachromatic granules is presumptive evidence for *C. diphtheriae*.

> Primary stain: Albert's stain
> Counterstain: Gram's iodine
>      or
> Primary stain: Loeffler's methylene blue
> Counterstain: None

The metachromatic-granule staining procedure uses a stain that adheres strongly to the phosphate crystals composing the granules. In the Albert's procedure, the smear is covered with the stain and allowed to react for a few minutes. It is then rinsed and counterstained with Gram's iodine. Cells with metachromatic granules will have dark blue to black granules in their cytoplasm, which appears uniformly light green.

The Loeffler's procedure requires simple staining with the methylene blue for several minutes. The granules appear dark blue against a light blue cytoplasm.

> Positive cells: Have visible granules in their cytoplasm

## LABORATORY OBJECTIVES

Differential and selective stains allow the microbiologist to separate bacteria rapidly into smaller, more manageable groups for identification. For this exercise you should:

1. Understand the significance of the gram-stain procedure and know what components

of bacterial cells are responsible for the differential staining reactions.
2. Understand the clinical importance of the acid-fast and metachromatic-granule staining procedures; know how these staining procedures work.
3. Be able to discuss the components of differential staining procedures and why they differ from simple staining procedures.
4. Begin to categorize bacteria on the basis of their cellular morphology and staining characteristics.

## MATERIALS NEEDED FOR THIS LABORATORY

1. Twenty-four-hour cultures of:

    *Staphylococcus epidermidis:* On nutrient agar slants
    *Escherichia coli:* On nutrient agar slants
    *Mycobacterium smegmatis:* On Lowenstein-Jensen medium
    *Corynebacterium xerose:* On Loeffler's serum slants
    *Branhamella catarrhalis:* On nutrient agar slants

2. Twenty-four-hour cultures of other bacteria studied in the previous exercises, as directed by your instructor.
3. Differential staining reagents:
    Gram-stain reagents:
        Crystal violet
        Gram's iodine
        Ethanol or acetone/ethanol
        Safranin
    Acid-fast reagents:
        Ziehl-Neelsen carbon fuchsin
        Acid alcohol
        Methylene blue
    Metachromatic-granule stain reagents:
        Albert's stain
        Loeffler's methylene Blue
4. Small pieces of paper toweling, approximately 3/4 in. by 2 in.
5. Prepared slides of *C. diphtheria* and *M. tuberculosis*.

## LABORATORY PROCEDURE

The aseptic techniques required in the previous exercise are also necessary here. Use the same precautions to prevent staining of your fingers, clothing, and jewelry.

1. Prepare smears of all the bacterial cultures. Make one set of smears for each staining procedure, as shown:

    Gram stain: All cultures
    Acid-fast stain: Any mycobacteria, and at least two negative controls
    Metachromatic granules: *Corynebacterium xerose* and at least two negative controls

2. When differential stains are used, proper controls are needed to ensure that the procedure is working correctly. An easy way to employ such controls is to make sure that there is a positive and negative smear on the same (why?) slide.
3. Complete gram stains on one set of smears:
    a. Flood the slides with crystal violet.
    b. After about 30 seconds, remove the crystal violet by rinsing gently with the iodine solution.
    c. Flood the smears with Gram's iodine and allow the iodine to react for about two minutes.
    d. As you hold the slide over the sink or staining tray, allow the ethanol or acetone/ethanol to flow over the smears. As soon as the color stops flowing from the smear, rinse the slide with water. This step, the *decolorization* step, is the one that seems to be most critical. It is easy to use too much solvent, resulting in poorly stained gram-positive cells.
    e. Cover the smears with safranin for three to five minutes.
    f. Rinse the stain from the slide, drain the excess water from the slide, and air dry.
4. Complete acid-fast stains on all smears of *Mycobacterium* and at least two negative controls.
    a. Construct a steam bath or use a ring stand to support the slide while it is being heated.

b. Cut or tear a small piece of paper toweling that is larger than the smears, but slightly smaller than the slide (about 3/4 in. wide by 2 in. long).
   c. Place the slide on the steam bath or ring stand, cover the smear with the piece of paper toweling, and flood the slide with Ziehl-Neelsen carbol fuchsin.
   d. Heat the slide for *not less than* five minutes. If you use a Bunsen flame to heat the slide you should heat it enough to cause vapors to appear over the stain. *Do not allow the fluid to boil and do not allow the stain to dry*. As the stain evaporates, add more
   e. After the smears have been heated for at least five minutes, allow the slides to cool, then rinse with water to remove excess stain. Do not allow the paper to fall into the sink. Use your needle, loop, or forceps to place it in the waste container.
   f. Rinse the slide with acid alcohol until stain no longer flows from the smear. The acid-fast stain, while not as rapidly decolorized as the gram stain, can be destained if the solvent is applied for too long a period of time. Rinse gently with water.
   g. Counterstain with methylene blue for three to five minutes, rinse, drain, and air dry.
5. Complete the metachromatic-granule stain on any smears of *Corynebacterium* and at least two negative controls.
   a. Flood the smears with Loeffler's methylene blue for three minutes.
   
   or
   
   a. Flood the smears with Albert's stain for three to five minutes.
   b. Rinse the slide with water and drain.
   c. Flood the smear with Gram's iodine and allow to react for about one minute.
   d. Rinse, drain, and air dry.
6. Examine all smears with the oil-immersion objective.
7. Examine the prepared slides as directed.
8. Complete the Laboratory Report Form and fill in the data on the Bacterial Characteristics Form. Consult the Text and Resource References for those staining characteristics that were not covered in this exercise.

## REFERENCES

### Text References

1. Atlas, Ronald M., *Microbiology: Fundamentals and Applications*. Chapters 2, 4, and 11.
2. Brock, Thomas D., David W. Smith, and Michael T. Madigan, *Biology of Microorganisms,* 4th ed. Chapters 2 and 18. Appendix 5.
3. Jensen, Marcus, *Introduction to Medical Microbiology*. Chapter 3.
4. Nester, Eugene W., et al., *Microbiology*, 3rd ed. Chapter 3.
5. Stanier, R. Y., E. A. Adelberg, and J. Ingraham, *The Microbial World,* 4th ed. Chapters 5 and 23.
6. Stanier, R. Y., et al., *Introduction to the Microbial World*. Chapter 3.
7. Wistreich, George A., and Max D. Lechtman, *Microbiology,* 4th ed. Chapters 4 and 9.

### Resource References

1. Finegold, Sidney M., and W. J. Martin, *Bailey and Scott's Diagnostic Microbiology,* 6th ed. Chapters 3 and 43.
2. Gerhardt, P., ed., *Manual of Methods for General Bacteriology*. Chapter 3.
3. Lennette, E. H., ed., *Manual of Clinical Microbiology*, 3rd ed. Chapter 98.
4. McGonagle, L. A., *Procedures for Diagnostic Microbiology*. Pages 119, 123, 127, 130.

## LABORATORY PROTOCOL

**Exercise 3—Differential Staining Procedures
(Gram Stain, Acid-Fast Stain, Metachromatic-Granule Stain)**

Check each step when you complete it.

**First Day**

_____ 1. Prepare smears of all cultures according to the following table:

        Gram stain:                    *All* cultures

        Acid-fast stain:             Any mycobacteria, and at least two negative controls

        Metachromatic granules:   *Corynebacterium xerose* and at least two negative controls

_____ 2. Complete all gram stains.

_____ 3. Complete all acid-fast stains. Remember to allow the slides to cool completely after the heating and to discard the paper in the appropriate container, *not the sink.*

_____ 4. Complete the metachromatic-granule stains.

_____ 5. Examine all smears with oil-immersion optics. If the stains did not come out correctly, check with the instructor to determine the cause.

_____ 6. Repeat any stains that did not come out correctly. It is essential that you learn to do these stains correctly.

_____ 7. Examine the prepared slides and report as directed.

_____ 8. Complete the Laboratory Report Form and one Bacterial Characteristics Form for each organism. Use the references to look up those staining characteristics not covered in this exercise.

# NOTES

Name: _____

## LABORATORY REPORT FORM

**Exercise 3—Differential Staining Procedures**
**(Gram Stain, Acid-Fast Stain, Metachromatic-Granule Stain)**

1. Complete one Bacterial Characteristics Form for each organism. These forms are provided in Appendix 6.

2. Write a short description of each of the staining procedures used in this exercise:

    Gram stain:

    Acid-fast stain:

    Metachromatic-granule stain:

## ANSWER THE FOLLOWING QUESTIONS

1. Define:

    Primary stain:

    Mordant:

Counterstain:




Decolorization:




2. List three diseases that are caused by bacteria that would be detected by the staining reactions studied in this exercise.

   a.


   b.


   c.


3. Why should a control smear be stained on the *same* slide as the unknown smear?




4. What would happen if you did not heat the acid-fast smear for a long enough time?




5. List all the morphological characteristics of *Corynebacterium diphtheriae*.

# Exercise 4
# Additional Differential Staining Procedures (Demonstration of Bacterial Cytology)

There will be four staining techniques demonstrated in this exercise. The exercise is structured so that each of them can be completed independent of the others. Your laboratory instructor will indicate which of the procedures you should complete.

The staining procedures covered in Exercise 3 (gram stain, acid-fast stain, and metachromatic-granule stain) are commonly used for the identification of clinical isolates. However, they are only three of many such procedures. The procedures that will be covered in this exercise, while not as commonly used in routine laboratory work, are useful because they stain cellular organelles that would not otherwise be visible. Their diagnostic value is somewhat less than the three other procedures, but their value for the study of bacterial cytology is greater.

The structures that are revealed by these procedures are, for the most part, subcellular organelles and are smaller than the cells that contain them. You will need to pay special attention to getting maximum resolution from your microscope. Perhaps you should review Chapter 1. Examination of prepared slides would be quite helpful for reviewing your microscope and staining techniques.

## NEGATIVE STAIN

The *negative stain,* as the name implies, stains the background but does not stain the cell itself. Its effect is not unlike that of a photographic negative: Everything appears dark except the structure you want to examine. The negative staining procedure is frequently used to examine the capsules of bacteria, but it can be used to stain virtually any structure that is impermeable to the stain. (We will discuss specific capsule stains in a later section of this exercise. Why not compare the two procedures?)

The most commonly used procedure is to mix the bacteria with india ink and examine the resulting suspension as a wet mount (Exercise 2) or as a very thin smear. India ink is a colloidal suspension of carbon particles. When bacteria are mixed with the india ink, the carbon particles cannot penetrate the surface layers of the cell, and the cell remains unstained while the background appears dark. We will use simple negative-staining procedures in this exercise. The Resource References contain other protocols, including some that stain the cell. In these latter procedures, the cell will assume the color of the stain, the capsule is unstained, and the background is black or dark brown.

## THE CAPSULE STAIN

The capsule or slime layer is a deposit of slippery, mucoid material that forms around almost all bacterial cells. Some forms produce capsules that are very small, while others, such as *Klebsiella pneumoniae* and *Streptococcus pneumoniae,* produce prominent capsules. Many studies have linked the capsule with enhanced bacterial virulence, and the capsule has played a fairly prominent role in the history of microbiology. Consult the Text References for information about capsules and bacterial transformation and the Quellung reaction.

Although it is possible to see the capsule with negatively stained smears, several techniques can be used to actually stain the capsules and/or the cells. In one procedure, a negatively stained thin smear is gently heated and then lightly stained. In another procedure, known as the Anthony method, the capsules are coated with copper sulfate to make them appear light blue.

## THE SPORE STAIN

Bacterial endospores, as explained in Chapter 3, are resistant to temperatures that would kill vegetative cells. By far the most commonly encountered endospore-forming bacteria belong to the genera *Bacillus* and *Clostridium,* although at least one coccus is known to make endospores.

The morphology and location of the spore, relative to the cell that produced it, is quite varied. Endospore morphology is an interesting study; the variations are used to help classify the bacilli and the clostridia. When a spore-forming bacterial species is described, the following characteristics of the spore are usually noted:

- Shape:
    Oval
    Round
    Relative length or diameter
- Size:
    Larger than the cell
    Smaller than the cell
    About the same diameter
- Location:
    Central (in the middle)
    Terminal (at the very end)
    Subterminal (in between)
    Lateral (off the long axis)

The wall of the endospore is impermeable to water—and to anything dissolved in the water. However, heating the cells in the presence of a stain allows sufficient dye to enter the spore so it can be seen. The procedure used is identical to that used to stain acid-fast bacteria (Exercise 3).

## THE FLAGELLA STAIN

Bacteria that produce flagella are motile. A positive test for motility by the wet-mount procedure or by the use of motility medium is evidence that the bacterium being tested has flagella. Unfortunately, however, simply knowing whether or not a bacterium has flagella is not enough.

In many protocols for the identification of the gram-negative rods, serological methods are used to test the antigenic nature of the flagella. In other cases it is important to determine the location and number of flagella on a single cell.

The location and sometimes the number of flagella on bacterial cells has been found to be consistent for any given species and may be used to help classify bacteria into their correct taxons. Consult the References to find out which bacteria have peritrichous and which show polar flagellation. If they are polarly flagellated, do they have one flagellum at one end, one at both ends, or a tuft of flagella at one or both ends?

Flagella, almost without exception, have diameters considerably smaller than the resolving power of the best optical instrument (even when it is used correctly). The flagella stain makes these organelles visible by coating them with sufficient amounts of mordant (tannic acid and alum) to make them visible. Two procedures are common: Gray's method and Leifson's method. Gray's method does not require special dyes, but it is somewhat more touchy in terms of getting excellent results. In any case, both procedures require that: (a) the slide be absolutely clean, (b) the cells be handled in an exceptionally gentle manner to avoid breaking off the flagella, and (c) capsular material be removed before staining.

## LABORATORY OBJECTIVES

1. Understand the nature of the bacterial endospore and its relationship to the development of sterilization procedures.
2. Understand the nature of the capsule and its relationship to virulence.
3. Understand the nature of bacterial flagella and their role as organelles of locomotion.

4. Gain additional experience with high-resolution microscopy and with additional cytological techniques.

## MATERIALS NEEDED FOR THIS LABORATORY

1. For the *negative-stain* procedure:
   a. Twenty-four-hour cultures of bacteria. Although encapsulated bacterial species are ideal for demonstrating negative staining, other bacteria work equally well. Representatives of the following genera usually produce good results:
      *Proteus*
      *Pseudomonas*
      *Streptococcus*
      *Bacillus*
   b. Good-quality india ink. Use india ink that has the smallest-size carbon particles. Most artist- and engineering-quality india ink is satisfactory.
   c. Glass slides and coverslips, as needed.
2. For the *capsule stain*:
   a. Twenty-four-hour cultures of capsule-forming bacteria. Select species that form prominent capsules:
      *Klebsiella pneumoniae*
      *Streptococcus pneumoniae*
   b. Good-quality india ink.
   c. Crystal violet: 1%, wt/vol, aqueous.
   d. Copper sulfate (with 5 waters of hydration); 20%, wt/vol, aqueous.
3. For the *endospore stain*:
   a. Seventy-two-hour cultures of spore-forming bacteria. Choose bacterial forms that present various spore morphologies.
      *Bacillus cereus*
      *Bacillus subtilis*
      *Bacillus megaterium*
      *Clostridium sporogenes*
   b. Several of the above species, as well as selected important pathogens, may be demonstrated with prepared slides.
      *Clostridium tetani*
      *Clostridium botulinum*
   c. Malachite green; 0.5%, wt/vol, aqueous.
   d. Gram's safranin
4. For the *flagella stain*:
   a. Twenty-four-hour cultures of motile bacteria. Choose bacteria that demonstrate peritrichous and polar flagellation.
      *Proteus vulgaris*
      *Pseudomonas aeruginosa*
   b. Carefully cleaned slides, or the material to clean them.
   c. Commercially prepared slides are available to serve as controls for staining technique and microscope adjustment.
   d. Tubes of water or nutrient broth as needed to suspend agar cultures.
   e. Gray's solution "A." This solution should be assembled from the stock solutions just before use. It does not store well and must be filtered just prior to use.
   f. Ziehl's carbon fuchsin.

## LABORATORY PROCEDURE

### Negative Staining Procedure (Wet Mount)

1. Use a bacterial strain that produces a large capsule. *Klebsiella pneumoniae* or *Streptococcus pneumoniae* are quite suitable and usually exhibit typical morphology and produce good capsules when grown on a carbohydrate-rich medium.
2. If the organisms are growing on an agar slant, transfer a loopful of culture into about 2 ml of saline. If they are in broth, you do not need to suspend them in saline.
3. Be especially careful that the slides are clean and free of grease. It is essential that the fluids (saline or broth and india ink) mix freely and flow smoothly over the surface of the slide.
4. Place one loopful of india ink on the slide. Place a loopful of the bacteria suspension next to, *but not touching,* the drop of india ink. *Do not mix the drops at this time.*
5. Carefully lower a coverslip over the drops. Allow it to cover both the india ink and the suspension of bacteria. When the coverslip is in place, the drops will have mixed in such a way that a stained gradient will be formed.

6. If a sealed wet mount is desired, place the droplets on the coverslip and use the procedure given in Exercise 2.
7. Examine the wet mount with the low-power and the high-dry objective lenses. Find the part of the slide that has the optimum contrast for observing both capsules and the refractile bacterial cell.
8. The wet-mount staining procedure does not subject the cell to drying. Any artifacts that occur when cells are dried will be avoided.

**Negative-Staining Procedure (Thin Smear)**

1. Use the same bacteria as before.
2. Place one loopful of india ink near the end of a glass slide. Mix a loopful of bacterial suspension with the india ink. Try to keep the droplet as small in diameter as possible.
3. Hold a short edge of another glass slide against the first slide, at an angle of about 45 degrees. Beginning near the center of the slide, "back" the tilted slide up until it touches the mixture of india ink and bacteria. At this point the mixture should spread out along the edge of the slide.
4. Quickly push the tilted slide forward until it clears the slide that the smear will be on. If the slide was clean and your pushing was smooth, the droplet will have smeared across the slide, producing a thin smear with a feathered edge. Allow the slide to air dry.
5. Observe the organisms with both the low-power and high-dry objective lenses. The best part of the slide is usually near the feathered edge. The cells should be distinctly visible as unstained structures in a smooth-textured, brown background.

**Capsule Stain (Thin Smear Procedure)**

1. Prepare a negatively stained thin smear, as described above.
2. Gently heat fix the smear.
3. Flood the smear with crystal violet for about one minute.
4. Gently wash the slide with cold water. Air dry.
5. Examine the slide with all three objective lenses. The cells should be stained dark blue to purple and should be approximately in the center of the unstained capsule. The background is, as before, brown to black.

**Capsule Stain (Anthony's Method)**

1. Prepare a smear as described in Exercise 2. Allow the smear to air dry. Do not heat fix.
2. Stain the smear with 1% aqueous crystal violet for about two minutes.
3. Gently wash the slide with a 20% solution of copper sulfate. Drain. Blot dry.
4. Examine the slide with all three objective lenses. The cells should appear dark blue or purple, with light blue capsules.

**Endospore-Staining Procedure (Schaeffer-Fulton Method)**

1. Seventy-two-hour cultures of *Bacillus* or *Clostridium* should be used.
2. Prepare a smear of the cultures in the usual manner. Air dry and heat fix.
3. Place a small piece of filter paper over the smear (refer to the acid-fast procedure in Exercise 3) and flood it with the malachite green stain.
4. Gently heat the flooded smear with a Bunsen flame or by steaming it over a boiling water bath. The stain should steam for at least five minutes. Do not allow the stain to boil; do not allow the smear to dry; add additional stain as needed.
5. After the slide has cooled, discard the paper in a suitable container and wash the slide with cold water until color no longer flows from the smear.
6. Counterstain with safranin for one to three minutes. Rinse with cold water and air dry.
7. Examine the slides with all three objective lenses. The spores will stain green and the cytoplasm will be pink.

**Flagella-Stain Procedure (Gray's Method)**

1. Use any motile organisms, including peritrichous and polar-trichous species. *Proteus vulgaris* and *Pseudomonas aeruginosa* usually give excellent results.

2. Sixteen-to-twenty-four-hour broth cultures produce excellent results and may be directly applied to the slide. If it is necessary to use an agar culture, transfer enough loopfuls of culture to produce a distinctly turbid suspension in nutrient broth or in saline and allow the cells to swim around for about 30 minutes. This will cause the bacteria to lose most of their capsular material.

3. Carefully clean one glass slide for each of the bacteria to be stained. A paste of nonabrasive cleanser (such as Bon Ami) can be rubbed on the glass and rinsed in flowing warm water. Either air dry or wipe with a fresh Kleenex or Kimwipe tissue. After the slide has been cleaned, do not touch the part of the slide that will hold the smear.

4. Place one drop of the bacterial suspension near the edge of the slide. Tilt the slide and allow the droplet to flow over the surface. Rock the slide to spread out the smear. Do not use your needle or loop to spread the droplet.

5. Allow the smear to air dry.

6. Flood the smear with filtered Solution A. Allow the cells to stain for about eight minutes. Some experimentation may be needed to determine the best staining time; between six and ten minutes usually works well. (Why not stain several slides for different times [6 through 10 minutes]?). Some technicians filter Solution A by letting it drip through filter paper (in a funnel) directly onto the slide.

7. Gently rinse the slide with distilled water. Do not squirt the water directly on the smear. Rather, allow the water to flow over it until the stain is removed.

8. Cover the smear with a small piece of paper toweling or blotting paper. Flood the smear with acid-fast (Ziehl's) carbol fuchsin for three minutes.

9. Remove the paper by lifting (not sliding) it off the smear. Rinse gently with distilled water. Air dry.

10. Examine the slide with all three objective lenses. You may need to search several sections of the slide to find good examples of the flagellated cells. The cells and their flagella will stain red.

## REFERENCES

### Text References

1. Atlas, Ronald M., *Microbiology: Fundamentals and Applications.* Chapters 2, 4, and 11.
2. Brock, Thomas D., David W. Smith, and Michael T. Madigan, *Biology of Microorganisms,* 4th ed. Chapter 2. Appendix 5.
3. Jensen, Marcus, *Introduction to Medical Microbiology.* Chapter 3.
4. Nester, Eugene W., et al., *Microbiology,* 3rd ed. Chapter 3.
5. Stanier, R. Y., E. A. Adelberg, and J. Ingraham, *The Microbial World,* 4th ed. Chapters 11 and 22.
6. Stanier, R. Y., et al., *Introduction to the Microbial World.* Chapters 6 and 12.
7. Wistreich, George A., and Max D. Lechtman, *Microbiology,* 4th ed. Chapters 4 and 9.

### Resource References

1. Finegold, Sidney M., and W. J. Martin, *Bailey and Scott's Diagnostic Microbiology,* 6th ed. Chapters 3 and 43.
2. Gerhardt, P., ed., *Manual of Methods for General Bacteriology.* Chapter 3.
3. Lennette, E. H., ed., *Manual of Clinical Microbiology,* 3rd ed. Chapters 13, 37, and 98.
4. McGonagle, L. A., *Procedures for Diagnostic Microbiology.* Pages 120-122, 129.

# LABORATORY PROTOCOL

**Exercise 4—Additional Differential Staining Procedures
(Negative Stain, Capsule Stain, Endospore Stain, and Flagella Stain)**

Check each step when you complete it.

**First Day**

_____ 1. Complete all negative stains, as directed by your instructor.

_____ 2. Complete all capsule stains, as directed by your instructor.

_____ 3. Compare the results obtained with the two types of capsule- and negative-staining procedures. Did you observe any differences in the size or shape of either the capsule or the cells?

_____ 4. Complete all endospore stains, as directed by your instructor.

_____ 5. Carefully compare the endospore morphology of each species studied. Record these observations on the Laboratory Report Form.

_____ 6. Complete all flagella stains, as directed by your instructor.

_____ 7. Compare the location and number of flagella observed on each of the species studied. Record the results on the Laboratory Report Form.

# NOTES

Name: _____

## LABORATORY REPORT FORM

**Exercise 4—Additional Differential Staining Procedures**
**(Negative Stain, Capsule Stain, Endospore Stain, and Flagella Stain)**

1. Complete one Bacterial Characteristics Form for each organism. You may wish to simply fill in or complete the forms used in Exercise 1 and 2. Use the "Notes" section of the form to record special morphological features of spores and flagella.

2. Write a short description of each of the staining procedures used in this exercise:

   Negative stain:




   Capsule stain:




   Endospore stain:




   Flagella stain:




## ANSWER THE FOLLOWING QUESTIONS

1. Define:

   Flagellum:

**64** EXERCISE 4

Endospore:

Capsule:

2. List three diseases that are caused by bacteria that form endospores.
    a.

    b.

    c.

3. What would you use as a control for each of the staining procedures used in this exercise?

4. What would happen if you did not heat the spore stain for a long enough time?

5. List all of the morphological characteristics of endospores.

6. List all of the morphological characteristics of flagella.

# Exercise 5
# Physiological Characteristics of Bacteria

The next three exercises cover the physiological characteristics of bacteria. The Laboratory Report Form for all three exercises (5–7) will be found at the end of Exercise 7.

## INTRODUCTION TO BIOCHEMICAL CHARACTERIZATION OF BACTERIA

There are at least two reasons for studying the biochemical characteristics of bacteria. First, the characteristics can be used to demonstrate the exceptional metabolic diversity of procaryotic organisms. The range of metabolic capabilities that bacteria exhibit is very large, explaining in part why these organisms can be found in virtually every environmental habitat. Not only are bacteria capable of obtaining energy by a variety of pathways, some of which are unique to bacteria, but they are capable of utilizing a very large number of different metabolites.

The second reason is that the biochemical characteristics of bacteria represent additional phenotypic characteristics that can be easily examined. The fact that individual species characteristics are genetically determined makes it possible to use them as phenotypic markers. These biochemical characteristics make it possible to identify unknowns by matching the phenotype of the unknown to that of a known reference organism. A reference organism is one that is considered to be typical of a given genus and species—so much so that it can be used as a reference against which unknown isolates can be compared.

## PART ONE: REACTIONS OF CARBOHYDRATES

The reactions of carbohydrates that are employed for identification are usually degradative catabolic reactions used by the bacteria as part of their energy-producing metabolism. The synthesis of structural and storage carbohydrates is also important, but these reactions are not often used in identification protocols. The reactions you will study will fall into two major categories: those which determine *if* certain carbohydrates are used at all, and those which determine *how* they are used. For example, we will test several bacterial species to determine which sugars they are able to use (i.e., glucose, lactose, etc.) and whether or not the sugars can be fermented with the production of acid and gas, or only acid.

## EXPERIMENTAL METHODS

In order to determine an organism's biochemical characteristics, you must use a medium that induces or enhances that characteristic and you must use some chemical test to measure the activity. The latter often involves the use of pH indicators in the medium to detect the production of either acids or bases, but may also depend upon added chemicals that react with the products to give a colored compound. Some detailed examples are given below.

**Fermentation of sugars.** Sugar-fermentation broths contain the sugar being tested and a pH indicator to detect the production of acid. In this exercise, you will use phenol red sugar broths. If the sugar is fermented, the medium will turn from red to yellow because of the production of acid. The glucose sugar tube will also have a smaller tube inside for the entrapment of gas bubbles, should any gas be produced. If the bacteria do not utilize the sugar, they will often use the amino acids contained in the medium. When amino acids are used, ammonia is produced as a by-product, causing the pH of the medium to rise (the color will turn deeper red). If the sugar is used oxidatively (respiration), there will be little, if any, color change. Sugar fermentation results are reported as *acid* (A), *acid + gas* (A+G), *alkaline* (Alk), or *neutral* (N). (Can you explain why?)

**Methyl red (M-R) test.** Methyl red is a pH indicator that changes color at a lower pH (about 4.5) than phenol red. When a few drops are added to a broth containing a sugar, a color change will result if the pH falls below about 4.5. You should remember that this indicator turns *red* below a pH of 4.5, and *yellow* in a pH above that point. Unlike the phenol red used in the fermentation broths, a red color is positive for the methyl-red test, indicating a pH on the acid side of this indicator. This test is used in most identification schemes for the gram-negative rods.

**Citrate utilization.** The medium (Simmons citrate agar) used to determine if a given isolate is able to utilize citrate has citrate as the only source of oxidizable carbohydrate. It also has a pH indicator that changes color if the citrate is used. The indicator (Brom thymol blue) turns from yellow to blue as the citrate is utilized. The reasons for the pH change, which is from neutral (a green color—why?) to alkaline, are complicated and related to the reactions of the alkaline earth metals in an aqueous medium. For the purposes of this exercise, it is necessary only to note that as the citrate is used, these reactions will occur and the indicator will change from green to blue.

## A NOTE ABOUT pH INDICATORS

pH indicators will change color as the pH of the solution they are in changes. The actual color change takes place over a range, the center of which is referred to as the pK. The indicators you will use in this exercise are:

| Indicator | pK | Range and Color |
|---|---|---|
| Methyl red | 5.2 | 4.4–6.0 Red-Yellow |
| Phenol red | 7.9 | 6.8–8.4 Yellow-Red |
| Brom thymol blue | 7.0 | 6.0–7.6 Yellow-Blue |

These data indicate that phenol red will be yellow below pH 6.8, that it will begin to turn red at 6.8, and that it will be completely changed by pH 8.4. Between 6.8 and 8.4 it will be various shades of orange, becoming increasingly more red as the pH increases and increasingly more yellow as the pH decreases. Brom thymol blue will change from yellow to green to blue as the pH rises from 6.0 to 7.6.

**Voges-Proskauer (V-P) test.** This is a specific test for metabolic intermediates produced when sugars are fermented to produce butylene glycol in addition to acids. Butylene glycol fermentation is one of the fermentation patterns exhibited by some gram-negative rods. It is an alternate fermentation pattern to those that result in the production of acids only. A positive V-P test is almost never observed with a positive M-R test because the M-R test is based on the production of sufficient acid to lower to pH to below 4.5. If some of the sugar being fermented is used to make butylene glycol, then it cannot be used to produce acids. In this test you add Barritt's reagents (A and B) to the medium and observe a color change after allowing sufficient time (about ten minutes) for the reaction to occur.

**Hydrolysis of starch.** When iodine and starch react, a deep purple color results. If a starch agar plate is flooded with an iodine solution (Gram's iodine may be used), any starch present will react with the iodine, causing a deep purple color to form. If the starch has been hydrolyzed around the colonies, the medium will remain unstained. The enzyme secreted by bacteria that hydrolyzes starch is called amylase; the test is occasionally referred to as the *amylase test*.

## OBJECTIVES OF PART ONE

The objectives of this exercise are related to the reasons given for the study of bacterial biochemical characteristics. You should:

1. Understand the difference between fermentation and respiration and why a pH change is indicative of fermentation.
2. Obtain an appreciation for the metabolic capabilities of bacteria and how these capabilities can be used to identify bacteria.
3. Become introduced to the characterization of bacteria on the basis of their phenotypic characteristics.

## MATERIALS NEEDED FOR THIS LABORATORY

*Note:* This material list is for *each* organism tested. Your instructor will indicate how many of the bacterial species each student is to test.

1. Culture tubes or Durham tubes containing phenol red sugar broths. (A Durham tube is one that has a small inverted tube immersed in the broth to catch any gas that might be produced.)
    Glucose broth (in a Durham tube)
    Sucrose broth
    Lactose broth
    Mannitol broth
2. One Petri plate with starch agar. (You may be able to test more than one bacterium on each plate. Check with your instructor.)
3. Two tubes of MR-VP medium, one for the methyl-red test and one for the Voges-Proskauer test.
4. One tube of Simmons citrate agar.
5. Reagents to complete the following tests:
    Methyl-red test: Methyl-red indicator in dropper bottles
    Voges-Proskauer test: Barritt's Reagents A and B in separate dropping bottles
6. Twenty-four-hour cultures of the organisms to be tested:
    *Staphylococcus aureus*
    *Staphylococcus epidermidis*
    *Streptococcus faecalis*
    *Escherichia coli*
    *Enterobacter aerogenes*
    *Proteus vulgaris*
    *Pseudomonas aeruginosa*
    *Bacillus subtilis*
    *Serratia marcescens*

## LABORATORY PROCEDURE

1. Obtain and label the following tubes or plates of media for each organism to be tested:
    Phenol red glucose broth (in a Durham tube)
    Phenol red lactose broth
    Phenol red sucrose broth
    Phenol red mannitol broth
    Simmons citrate agar slant
    MR-VP broth (two for each organism)
    Starch agar plate
2. Inoculate all media according to these instructions:
    a. Broths: Inoculate with either a loopful of culture or a needle. It is also possible to use sterile Pasteur pipettes, inoculating each tube with one drop of culture.
    b. Slants: Inoculate the surface of the Simmons citrate slant with a loop, a needle, or by allowing one drop from a sterile Pasteur pipette to flow over the surface. It is not necessary to stab into the citrate agar.
    c. Plates: Since isolation streaks are not needed at this time, a single streak in the center of plate will be sufficient. If you wish (depending upon the instructions you are given), you may use a single plate for up to four or five different bacteria. Inoculation streaks should be radiating from the center, spokewise, around the plate.
3. Incubate all cultures for 24 hours at 37°C.
4. Record the reactions observed in the sugar tubes and the citrate agar.
    *Note:* Detailed instructions for the preparation and use of the test reagents are given in Appendix 3.
5. Flood the starch agar plate with an iodine solution. Gram's iodine is the most convenient one to use. Allow several minutes for the color to develop. A clear (not stained purple) area around the colonies indicates that starch has been hydrolyzed. If the purple color extends up to the colony, no hydrolysis occurred.

6. Add about 0.5 ml of methyl-red indicator to one of the MR-VP broth tubes. If the indicator remains red, the test is positive; if it turns yellow, the test is negative.
7. Add 0.6 ml (about 12 drops) of Barritt's reagent A and 0.2 ml (about 4 drops) of Barritt's reagent B to 1.0 ml of broth from the second MR-VP culture. Mix well and wait ten minutes for the color to develop. A red color indicates a positive test, and a yellow color indicates a negative test.
8. Record results on the Bacterial Characteristics Form at the end of the exercise.

## REFERENCES

### Text References

1. Atlas, Ronald M., *Microbiology: Fundamentals and Applications.* Chapters 3, 5, 6, and 11.
2. Brock, Thomas D., David W. Smith, and Michael T. Madigan, *Biology of Microorganisms,* 4th ed. Chapters 5 and 18.
3. Jensen, Marcus, *Introduction to Medical Microbiology.* Chapter 4.
4. Nester, Eugene W., et al., *Microbiology,* 3rd ed. Chapters 6, 11, and 13. Appendixes IV and V.
5. Stanier, R. Y., E. A. Adelberg, and J. Ingraham, *The Microbial World,* 4th ed. Chapters 14 and 16.
6. Stanier, R. Y., et al., *Introduction to the Microbial World.* Chapters 8 and 9.
7. Wistreich, George A., and Max D. Lechtman, *Microbiology,* 4th ed. Chapters 5, 7, and 27.

### Resource References

1. Finegold, Sidney M., and W. J. Martin, *Bailey and Scott's Diagnostic Microbiology,* 6th ed. Chapters 7-32, and 44 (specific references).
2. Gerhardt, P., ed., *Manual of Clinical Microbiology,* 3rd ed. Chapters 8 and 20.
3. Lennette, E. H., ed., *Manual of Clinical Microbiology,* 3rd ed. Chapters 97 and 98.
4. McGonagle, L. A., *Procedures for Diagnostic Microbiology.* Consult specific topical references.

# LABORATORY PROTOCOL

## Exercise 5 — Physiological Characteristics of Bacteria (Part One: Reactions of Carbohydrates)

Check each step when you complete it.

### First Day

_____ 1. Obtain and label a complete set of media for *each* organism being studied. Each set should include:

| | |
|---|---|
| Phenol red glucose broth: | One |
| Phenol red lactose broth: | One |
| Phenol red sucrose broth: | One |
| Phenol red mannitol broth: | One |
| MR-VP broth: | Two |
| Citrate agar slant: | One |
| Starch agar plate: | Enough to test up to four bacterial species per plate |

_____ 2. Inoculate one full set of media with *each* of the test organisms provided. These may include:

*Staphylococcus aureus*
*Staphylococcus epidermidis*
*Streptococcus faecalis*
*Escherichia coli*
*Enterobacter aerogenes*
*Proteus vulgaris*
*Pseudomonas aeruginosa*
*Bacillus subtilis*
*Serratia marcenscens*

Refer to the Laboratory Procedure section of this exercise for instructions on how to inoculate each of these media.

_____ 3. Incubate all tubes and plates for 24 hours.

### Second Day

_____ 4. Record the results of the sugar fermentations. Note the color of the medium and whether or not gas bubbles are seen in the inner tube. Compare the color with an uninoculated control tube for each sugar.

*Note:* Detailed instructions for the preparation and use of test reagents are given in Appendix 3.

_____ 5. Flood the starch agar plates with an iodine solution and observe the color of the medium. If there is no purple color around the colonies, starch has been hydrolyzed.

_____ 6. Add several drops of methyl-red indicator to *one* tube of MR-VP broth. If the indicator stays red, it is a positive test; if it turns yellow, it is a negative test.

_____ 7. Add about 12 drops of Barritt's reagent A and about 5 drops of Barritt's reagent B to 1.0 ml of broth from the second tube of MR-VP medium. Mix well by swirling the tube. Allow at least 10 minutes for the color to develop. If a red color develops, the Voges-Proskauer test is positive.

8. Observe the color of the citrate agar slant. If the slant has turned blue, the citrate has been utilized. In a negative test, the medium remains green.

9. Record all results on the Biochemical Characteristics Form. Complete the Laboratory Report Form for Exercises 5, 6, and 7.

# NOTES

# Exercise 6
# Physiological Characteristics of Bacteria

Exercises 5, 6, and 7 cover the physiological characteristics of bacteria. The Laboratory Report Form will be found at the end of Exercise 7.

## PART TWO: REACTIONS OF PROTEINS

When there is no oxidizable carbohydrate in the medium, or if the bacteria are unable to use the one that is there, the organism must utilize any available amino acids and proteins, oxidizing the carbon chains and often excreting the unused, residual parts of the molecules, such as ammonia and hydrogen sulfide. All the reactions we will study are ones used by the bacteria when it is necessary for them to use proteins or amino acids as nutrients for energy metabolism.

## EXPERIMENTAL METHODS

Bacteria do not normally use proteins and amino acids as sources of carbohydrates, but prefer to save them for synthesis and growth (catabolite repression). We must force their use by withholding carbohydrate. Under such conditions, the bacteria will use proteins and amino acids; we can then test for the by-products of their decomposition or look for other evidence of proteolytic activity. Some examples are given below.

**Hydrolysis of gelatin.** Gelatin is a protein colloid, maintaining its gel state at temperatures below about 25°C. The gel colloidal state is also dependent on the structural integrity of the proteins found in gelatin, and any hydrolysis of the peptides will result in destruction of the colloid. Failure of the gelatin to solidify when it has been cooled is indicative of the hydrolysis of these proteins. To test for gelatin liquefaction, it is only necessary to inoculate a tube of nutrient gelatin, incubate for 24 hours, and then determine if the gelatin will solidify when cooled. Failure to solidify (when an uninoculated control has solidified) is a positive test for gelatin hydrolysis.

**Indole production.** *Indole* is a compound produced when the amino acid tryptophan is hydrolyzed to pyruvic acid (which is used for energy metabolism) and indol (which is excreted). The indole reacts with Kovac's reagent to produce a red color. The reagent does not mix with the aqueous medium and forms a layer on the surface. The red color develops in that layer.

**Hydrogen sulfide production.** Hydrogen sulfide is produced when bacteria utilize the sulfur-containing amino acids for energy metabolism. The hydrogen sulfide is excreted and can be tested for by taking advantage of the formation of a black precipitate when it reacts with either silver, iron, or lead (why does silver tarnish; why did the old lead-based white paints tend to turn dark near the seashore?) in the medium. To test for hydrogen sulfide, you must use a medium containing the sulfur-containing amino acids and a source of one of the metals mentioned above. Iron (as ferrous sulfate) or lead (as lead acetate) are the most commonly used sources. The medium is inoculated as a stab into an agar deep. When the hydrogen sulfide is produced, a black precipitate forms along the stab line.

**Urea hydrolysis.** Urea is produced as a by-product of protein and nucleic acid decomposition. Some bacteria are able to hydrolyze it to ammonia and carbon dioxide. The ammonia reacts with the water in the medium to produce ammonium hydroxide, causing the pH to rise. Urea medium contains phenol red indicator (what are its pK, range, and colors?) in addition to urea, and has an initial pH of about 6.5 to 6.8. Consult Exercise 5 to determine how the medium will change color as the ammonia is produced!

**Nitrate reduction.** Nitrate reduction is a characteristic shown by several organisms; it is frequently used in diagnostic protocol for gram-negative rods. Nitrate is reduced to nitrite, and chemicals are used to detect the nitrite. As described in Appendix 3, the test can be used to detect nitrite as well as any remaining nitrate and other reduction products of the nitrate. After nitrite is produced, it is, in turn, further reduced either to several gaseous products, including nitrogen gas, or to ammonia. The gaseous products are given off to the atmosphere, and the ammonia is frequently used for the synthesis of proteins, although it may be excreted into the medium under some circumstances.

## OBJECTIVES OF PART TWO

The objectives of this exercise are similar to those of Exercise 5 (Part One). Here, however, we can begin to look at the relationships between different pathways and functions within the cell. You should:

1. Understand catabolite repression and its relationship to the often-observed laboratory phenomenon of *reversion*.
2. Understand why some parts of amino acid molecules are excreted when the amino acids are used for energy metabolism.

## MATERIALS NEEDED FOR THIS LABORATORY

*Note:* This material list is for *each* organism tested. Your instructor will indicate how many bacteria each student is to test.

1. One each of the following tubes of media:
   Nutrient gelatin
   Tryptone broth
   Lead acetate agar
   Urea medium
   Nitrate agar slants
2. The following reagents will be required on the second day:
   Kovac's reagent (indole production)
   Nitrate reagents: Solutions A and B
3. Twenty-four-hour cultures of the organisms to be tested:
   *Staphylococcus aureus*
   *Staphylococcus epidermidis*
   *Streptococcus faecalis*
   *Escherichia coli*
   *Enterobacter aerogenes*
   *Proteus vulgaris*
   *Pseudomonas aeruginosa*
   *Bacillus subtilis*
   *Serratia marcescens*

## LABORATORY PROCEDURE

1. Obtain and label one complete set of media for *each* organism. Each set should include one tube of:
   Nutrient gelatin
   Tryptone broth
   Lead acetate agar
   Urea broth or agar
   Nitrate agar slant
2. Inoculate all media according to the instructions given below:
   Broths: Inoculate with a loop, needle, or Pasteur pipette.
   Agar stabs: The lead acetate and urea agar tubes are inoculated by stabbing with a needle into the center of the tube. A single stab is necessary to properly interpret some of the results.
3. Incubate all cultures for 24 hours.
4. Place the nutrient gelatin cultures, along with an uninoculated control, in a cold-water or ice bath. After the control has solidified, tilt each culture to determine if the gelatin has remained liquid. Any cultures that fail to solidify should be recorded as gelatinase positive.
   *Note:* Detailed instructions for the preparation and use of the test reagents are given in Appendix 3.

5. Add about 10 drops of Kovac's reagent to the tryptone broth. The reagent will form a layer on the surface and develop a red color if indole is present. The development of a red color should be recorded as positive for the production of indole.
6. Examine the lead acetate medium. Any blackening of the medium indicates that hydrogen sulfide has been produced.
7. Examine the tubes of urea medium. A red color indicates that the urea has been hydrolyzed and ammonia excreted. A red color should be recorded as a positive test.
8. Add several drops of nitrate reagent A, then nitrate reagent B to the tube of nitrate medium. If nitrite is present, a distinct red color will develop and the organism can be scored as nitrate positive.
9. If you used nitrate broth and a red color did not develop, add a small amount of zinc dust to the medium. If a red color develops after the addition of the zinc, it indicates that *nitrate, not nitrite,* was still present and the test was negative.
10. If no color develops in the broth when you add the zinc dust, then neither nitrate nor nitrite is present. The organism converted all the nitrate to a gaseous product and should be considered nitrate positive.
11. Complete all Laboratory Report Forms and fill in the Biochemical Characteristics Forms for these organisms. The Laboratory Report Form for Exercises 5, 6, and 7 is found at the end of Exercise 7.

## REFERENCES

### Text References

1. Atlas, Ronald M., *Microbiology: Fundamentals and Applications.* Chapters 3, 5, 6, and 11.
2. Brock, Thomas D., David W. Smith, and Michael T. Madigan, *Biology of Microorganisms,* 4th ed. Chapters 5 and 18.
3. Jensen, Marcus, *Introduction to Medical Microbiology.* Chapter 4.
4. Nester, Eugene W., et al., *Microbiology,* 3rd ed. Chapters 6, 11, and 13. Appendixes IV and V.
5. Stanier, R. Y., E. A. Adelberg, and J. Ingraham, *The Microbial World,* 4th ed. Chapters 14 and 16.
6. Stanier, R. Y., et al., *Introduction to the Microbial World.* Chapters 8 and 9.
7. Wistreich, George A., and Max D. Lechtman, *Microbiology,* 4th ed. Chapters 5, 7, and 27.

### Resource References

1. Finegold, Sidney M., and W. J. Martin, *Bailey and Scott's Diagnostic Microbiology,* 6th ed. Chapters 7-32, and 44 (specific references).
2. Gerhardt, P., ed., *Manual of Methods for General Bacteriology.* Chapters 8 and 20.
3. Lennette, E. H., ed., *Manual of Clinical Microbiology,* 3rd ed. Chapters 97 and 98.
4. McGonagle, L. A., *Procedures for Diagnostic Microbiology.* Consult specific topical references.

# LABORATORY PROTOCOL

## Exercise 6 — Physiological Characteristics of Bacteria (Part Two: Reactions of Proteins)

Check each step when you complete it.

**First Day**

_____ 1. Obtain and label a complete set of media for *each* organism being studied. Each set should include one tube of:

>Nutrient gelatin
>Tryptone broth
>Lead acetate agar
>Urea medium
>Nitrate medium (agar or broth)

_____ 2. Inoculate one full set of media with each of the test organisms to be tested. These may include:

>*Staphylococcus aureus*
>*Staphylococcus epidermidis*
>*Streptococcus faecalis*
>*Escherichia coli*
>*Enterobacter aerogenes*
>*Proteus vulgaris*
>*Pseudomonas aeruginosa*
>*Bacillus subtilis*
>*Serratia marcescens*

Refer to the Laboratory Procedure section of this exercise for instructions on how to inoculate each of these media.

_____ 3. Incubate all tubes and plates for 24 hours.

_____ 4. Set up any other test procedures suggested by your instructor.

**Second Day**

*Note:* Detailed instructions for the preparation and use of the test reagents are given in Appendix 3.

_____ 5. Complete all the test procedures and record the results on the Laboratory Report Form. Refer to the Laboratory Procedure section for instructions on how to complete these tests.

>_____ : Gelatin liquefaction
>
>_____ : Indole production
>
>_____ : Hydrogen sulfide production
>
>_____ : Urea hydrolysis
>
>_____ : Nitrate reduction (including the use of zinc dust if you used nitrate broth)

_____ 6. Fill in the appropriate spaces on the Biochemical Characteristics Form for each organism. The Laboratory Report Form for Exercises 5, 6, and 7 is found at the end of Exercise 7.

# NOTES

# Exercise 7
# Physiological Characteristics of Bacteria

## PART THREE: MISCELLANEOUS REACTIONS

Many bacteria produce enzymes that are secreted into the medium. For example, the hydrolysis of starch (Exercise 5) and the liquefaction of gelatin (Exercise 6) occur because some bacteria secrete hydrolytic enzymes that can depolymerize these molecules. Many of the excreted enzymes have ecological or medical significance, and a few have diagnostic significance as well.

The clinical significance of some of these enzymes lies in their ability to attack substrate in host cell membranes, tissues, or body fluids, thereby interfering with host defense mechanisms or homeostasis. In other instances the enzymes are antigenic and can be tested for by serological reactions that provide evidence of exposure to the bacteria that produced the enzymes.

The enzymatic reactions we will examine in this exercise are ones that have been shown to be useful in the identification of unknown isolates. They are, of course, also important for the normal metabolism of the cell.

When bacteria produce flagella they are able to swim through liquids and soft media. They are said to be motile. Motility, as we will test for it here, is the result of the activity of flagella by those species of bacteria that produce them.

It is usually possible to test for enzymatic reactions by adding the enzyme's substrate to the bacterial culture and then testing for the appearance of the product or the disappearance of the substrate. For example, when you tested for starch hydrolysis in Exercise 4, you grew the bacteria on starch agar and then added iodine to test for the hydrolysis of the starch.

## SPECIFIC ENZYMATIC AND MOTILITY TESTS

**Catalase test.** Catalase is an enzyme produced by many organisms, including man. Indeed, it is so commonly produced that those organisms that do not do so are rare enough so that the lack of catalase is a significant diagnostic characteristic. The enzyme converts hydrogen peroxide into water and oxygen and does so vigorously enough to cause foaming. (What happens when you use hydrogen peroxide to cleanse a cut?) Hydrogen peroxide is produced when some amino acids are directly oxidized (amino acid oxidase). It must be rapidly removed because it is highly toxic to most cells— hence, most cells that are aerobic or facultative produce catalase, and those cells that do not do so are either obligate anaerobes or microaerophilic. Catalase is tested for by dropping one drop of hydrogen peroxide (the substrate) on a bacterial colony. If bubbles appear (the product), the organism is catalase positive. The catalase test is a very important one for differentiating between the genera *Staphylococcus* and *Streptococcus*.

**Oxidase test.** Oxidase is an enzyme involved in certain oxidation reactions in aerobic bacteria. A relatively few genera, notably *Neisseria, Pseudomonas,* and a few other gram-negative rods produce it, making it an important diagnostic test. To perform the test, it is necessary to place a drop of the oxidase reagent (the substrate) on the colony being tested. If the colony is positive, it will turn first pink, then purple, and eventually black (the colored products are the result of the enzymatic activity). It is sometimes more convenient to simply flood the plate with the reagent and observe the color changes. The oxidase reagent is, however, toxic, and positive colonies must be rapidly transferred if they need to be subcultured.

**Coagulase test.** Coagulase is an enzyme produced by *Staphylococcus aureus,* which coagulates blood plasma. The enzyme is *not* produced by *S. epidermidis* or by any other gram-positive coccus. This test is very important because it allows for rapid identification of an important and commonly encountered pathogenic bacterium. A rapid screening test for coagulase may be completed by mixing a small amount of culture material with a drop of human or rabbit plasma. Try to emulsify the suspension. If the cells appear to agglutinate (clotting of the plasma), the test is positive. (What are the substrate and product for this enzyme?) Cells that are coagulase negative will produce an even, nonagglutinated suspension. A more definitive and sensitive test requires incubation of the bacterium in plasma at 37°C. The details of this test are given in Appendix 3.

**DNAase test.** DNAase is an excreted enzyme (like the gelatin and starch enzymes) that hydrolyzes DNA. A special medium, containing DNA and a dye, is used to measure this activity. The bacterial species to be tested is streaked on the surface of the plate and the culture incubated for 24 hours. If DNA is hydrolyzed (by DNAase), the medium will change color. DNAase production is an important diagnostic characteristic for the staphylococci and some gram-negative rods.

**Esculin hydrolysis.** Esculin hydrolysis (and growth in a medium containing bile) is considered to be definitive for the Enterococci, a group of streptococci that are frequently encountered in clinical samples. This test uses a medium containing both esculin and bile. Positive organisms (Enterococci) will grow and turn the medium black. The black color is due to a hydrolysis product of esculin. Bile is added as an additional test for the enterococci. It has no effect on the hydrolysis of esculin, except that it limits the type of organism that can grow on the medium (it must be resistant to bile).

**Motility.** Bacteria that have flagella are able to swim through liquids and soft media (less than one-half the amount of agar that is usually found in plated or slanted media). This ability can be observed directly with a wet-mount slide, or indirectly by using soft agar deeps. (Refer to Exercise 2 for instructions for the preparation of wet mounts). Soft agar deeps are inoculated with a single stab down the center of the deep. Motile bacteria will move outward from the stab, resulting in a medium with a very hazy and cloudy appearance. Nonmotile bacteria cannot move and therefore grow only along the stab, producing a clear and sharp stab line. Some media contain a dye that is turned red by bacterial action, making the above observations much more distinct. When you examine your results, the difference will be clear and obvious.

## OBJECTIVES OF PART THREE

The enzymes secreted by bacteria have both ecological and clinical significance. The production of these enzymes may also be of diagnostic significance. You should:

1. Try to understand the ecological significance of excreted enzymes. For example, how do bacteria and other microbes cause the decomposition of dead plant and animal remains?
2. Understand the possible clinical significance of certain bacterial enzymes (hemolysin, coagulase, collagenase, etc.). Those enzymes that are clinically significant may attack certain cells or inhibit certain processes in the host, or they may be serologically important.
3. Appreciate the role that certain enzymes play in the metabolism and in the identification of bacteria.

## MATERIALS NEEDED FOR THIS LABORATORY

*Note:* This materials list is for *each* organism tested. Your instructor will indicate how many bacteria each student is to test.

1. One each of the following tubes of media:
   Bile-esculin agar
   Motility medium
2. Sufficient nutrient agar plates for streaking out the test organisms. You may be able to streak more than one organism on each plate. However, do not try more than four, and remember, you will use these for the catalase and oxidase tests.
3. Sufficient plates of DNAase agar, with indicator, to test each of the organisms. As before, you should not try to test more than four bacteria on a single 100-mm-diameter petri plate.
4. The following reagents will be required on the second day:
   Hydrogen peroxide (catalase test)
   Oxidase reagent (oxidase test)
   Human or rabbit plasma
      (coagulase test)
5. Twenty-four-hour cultures of the organisms to be tested:
   *Staphylococcus aureus*
   *Staphylococcus epidermidis*
   *Streptococcus faecalis*
      (or other *Enterococcus*)
   *Escherichia coli*
   *Enterobacter aerogenes*
   *Proteus vulgaris*
   *Pseudomonas aeruginosa*
   *Bacillus subtilis*
   *Serratia marcescens*

## LABORATORY PROCEDURE

1. Obtain and label one complete set of media for *each* organism.
2. Inoculate all media according to the following instructions:
   a. Agar stabs: The motility agar tubes are inoculated by making a single stab with a needle into the center of the tube.
   b. Agar slants: Inoculate the surface of the slant with a loop or needle, or by allowing a drop of culture to flow over its surface.
   c. Plated agar: The plates should be inoculated by making three or four parallel streaks or four radiating streaks.
3. Incubate all cultures for 24 hours.

*Note:* Detailed instructions for the preparation and use of the test reagents are given in Appendix 3.

4. Observe the growth along the stab line in the motility medium. If the organism is nonmotile, the stab will be clearly defined; if the organism is motile, the stab will be cloudy and hazy, indicating that the bacteria have moved through the medium.
5. The presence of a black color in the bile-esculin agar indicates that esculin has been hydrolyzed. Also observe which species of bacteria were able to grow in the presence of bile.
6. Examine the color surrounding any growth on the DNAase test agar. The color change will be specific for the type of DNAase agar used, but in general, *any* color change indicates DNA hydrolysis.
7. Place one drop of hydrogen peroxide on each streak. If bubbles appear, the organism is recorded as catalase positive.
8. Place one drop of the oxidase reagent on another part of each streak, and observe the color changes over a five- to ten-minute period. Oxidase-positive colonies will turn pink almost immediately and then will gradually turn purple and black. Oxidase-negative colonies will not turn color.
9. The coagulase test need only be run on *Staphylococcus aureus* and *Staphylococcus epidermidis.* Place one drop of human or rabbit plasma on a glass slide. Using your inoculating loop, transfer a small amount of culture and mix it with the drop of plasma. A coagulase-positive organism will appear to clump and agglutinate, while one that is coagulase negative will produce an even suspension with no clumping. Your instructor may want you to complete the more sensitive test that is shown in Appendix 3.
10. Complete the Laboratory Report Form and one Biochemical Characteristics Form for each organism studied.

## REFERENCES

### Text References

1. Atlas, Ronald M., *Microbiology: Fundamentals and Applications.* Chapters 3, 5, 6, and 11.
2. Brock, Thomas D., David W. Smith, and Michael T.

Madigan, *Biology of Microorganisms,* 4th ed. Chapters 5 and 18.
3. Jensen, Marcus, *Introduction to Medical Microbiology.* Chapter 4.
4. Nester, Eugene W., et al., *Microbiology,* 3rd ed. Chapters 6, 11, and 13. Appendixes IV and V.
5. Stanier, R. Y., E. A. Adelberg, and J. Ingraham, *The Microbial World,* 4th ed. Chapters 14 and 16.
6. Stanier, R. Y., et al., *Introduction to the Microbial World.* Chapters 8 and 9.
7. Wistreich, George A., and Max D. Lechtman, *Microbiology,* 4th ed. Chapters 5, 7, and 27.

## Resource References

1. Finegold, Sidney M., and W. J. Martin, *Bailey and Scott's Diagnostic Microbiology,* 6th ed. Chapters 7–32, and 44 (specific references).
2. Gerhardt, P., ed., *Manual of Methods for General Bacteriology.* Chapters 8 and 20.
3. Lennette, E.H., ed., *Manual of Clinical Microbiology,* 3rd ed. Chapters 97 and 98.
4. McGonagle, L. A., *Procedures for Diagnostic Microbiology.* Consult specific topical references.

# LABORATORY PROTOCOL

## Exercise 7—Physiological Characteristics of Bacteria (Part Three: Miscellaneous Reactions)

Check each step when you complete it.

### First Day

_____ 1. Obtain and label a complete set of media for *each* organism being studied. Each set should include:

| | |
|---|---|
| Bile-esculin agar | One tube |
| Motility medium | One tube |
| Nutrient agar plates | One plate |
| DNAase test agar | One plate |

(More than one organism, but not more than four, may be inoculated on each plate.)

_____ 2. Inoculate one full set of media with *each* of the test organisms to be tested. These may include:

*Staphylococcus aureus*
*Staphylococcus epidermidis*
*Streptococcus faecalis*
*Escherichia coli*
*Enterobacter aerogenes*
*Proteus vulgaris*
*Pseudomonas aeruginosa*
*Bacillus subtilis*
*Serratia marcescens*

Refer to the Laboratory Procedure section of this exercise for instructions on how to inoculate each of these media.

_____ 3. Incubate all tubes and plates for 24 hours.

_____ 4. Complete any other test procedures suggested by your instructor.

### Second Day

*Note:* Detailed instructions on the preparation and use of these test reagents can be found in Appendix 3.

_____ 5. Complete all the test procedures and record the results on the Laboratory Report Form. Refer to the Laboratory Procedure section for instructions on how to complete these tests.

_____ : Motility

_____ : Catalase production

_____ : Oxidase production

_____ : Coagulase (*Staphylococcus* only)

_____ : Esculin hydrolysis

_____ : Growth in presence of bile salts

_____ : DNA hydrolysis

# NOTES

Name: _____

## LABORATORY REPORT FORM

**Exercises 5, 6, and 7—Physiological Characteristics of Bacteria (Parts One, Two, and Three)**

1. Complete one Biochemical Characteristics Form for each organism studied in these exercises. The forms are provided in Appendix 5. If you do not have all the data, obtain the data you need from others in the laboratory or look it up in the Resource References. You will need it for Exercises 9 and 10.

### ANSWER THE FOLLOWING QUESTIONS

1. List two reasons for studying biochemical characteristics of bacteria.

    a.

    b.

2. Define *phenotype*.

3. Explain how alkaline reactions can occur in sugar fermentation tubes.

4. What is a pH indicator? Give two examples.

5. What is *reversion*? What is *catabolite repression*?

6. What is *hydrolysis*? List three examples.

7. Were there any bacteria that showed positive test results for all tests except the coagulase test?

8. List the substrate and product for each of the enzymes studied in Exercise 7.

9. Explain how some bacterial enzymes may have ecological significance.

10. Explain the medical significance of DNAase, coagulase, and catalase.

# Exercise 8
# Growth of Anaerobic Bacteria

A significant number of pathogenic bacteria are anaerobic. In addition to the well-known species of *Clostridium* (*C. tetani, C. botulinum, C. histolyticum,* etc.), there are many anaerobic and microaerophilic rods and cocci that are frequently encountered in clinical samples.

Cultures of the peritoneal cavity and of wounds are perhaps the most likely sources of anaerobic bacteria, although as anaerobic techniques have become more and more reliable, anaerobic bacteria are increasingly being found to be involved in pathogenic conditions. The culture and isolation of anaerobic bacteria involve enrichment in a semi-fluid medium followed by streak-plate isolation under anaerobic conditions.

## AEROBIC AND ANAEROBIC BACTERIA

Bacteria may be classified into four groups according to their requirements for molecular oxygen. In this exercise, these categories will be used to describe the species of bacteria being studied and not to describe a particular type of metabolism or reaction. As we will see, bacterial species that use anaerobic metabolic pathways may be called aerobes if they grow in the presence of oxygen (even though they may not use the oxygen). These four groups are:

**Aerobic.** A bacterial species that is *aerobic* is one that is able to grow in the presence of molecular oxygen. If the species *requires* molecular oxygen for metabolic energy production, it may be referred to as an *obligate aerobe*; it cannot grow unless molecular oxygen is present in the environment. Such bacteria usually do not ferment sugars, but oxidize them by a respiratory pathway. If, on the other hand, the bacterial species does not require oxygen, and is not inhibited by it, it will be able to grow whether or not oxygen is present.

**Anaerobic.** An anaerobic bacterial species is one that cannot grow in the presence of free oxygen. Oxygen is toxic to these bacteria; they will only grow in environments that are free of oxygen. In some instances, very small amounts of oxygen are toxic, and the requirement for an oxygen-free environment is absolute. Such bacterial species are often referred to as *obligate anaerobes*. Bacteria that do tolerate small amounts of oxygen are referred to as *microaerophilic*; they grow best in an environment that has little or no oxygen. In a way, the difference between *obligate anaerobe* and *microaerophilic* is one of degree. Both groups may be considered to be sensitive to oxygen, but one is much more so than the other.

**Facultative.** Bacteria that are able to grow without regard to the oxygen content of the environment are referred to as *facultative* organisms. They are not inhibited by oxygen, nor do they require it for their metabolism. There are at least two conditions that make an organism facultative. These include:

1. Bacteria that are able to use either an aerobic (aerobic respiration) or anaerobic (anaerobic respiration and/or fermentation) energy-producing metabolism. These bacteria simply (not

so simply) switch from one pathway to another, depending upon whether or not oxygen is present in the environment.
2. Bacteria that use anaerobic metabolic (fermentation) pathways, but which are not inhibited by oxygen. These bacteria will grow in the presence of oxygen, but do not use it.

For the purposes of this exercise, species of bacteria may be defined according to the following:

**Aerobic:** A bacterial species that grows in the presence of oxygen.
**Anaerobic:** A bacterial species that cannot grow in the presence of oxygen. It can be assumed that oxygen is toxic to the cell.
**Microaerophilic:** A bacterial species that grows poorly (but does grow) in the presence of oxygen. These bacteria grow better in oxygen-poor or oxygen-free environments.
**Facultative:** A bacterial species that grows well under both aerobic and anaerobic conditions. If this word is used as an adjective (as "facultative aerobe" and "facultative anaerobe"), confusing redundancy may be created. How would you differentiate between a "facultative aerobe" and a "facultative anaerobe"? Choose your words carefully.
**Obligate:** The term *obligate* may be used as an adjective to imply an absolute requirement for the environmental condition (as "obligate aerobe" and "obligate anaerobe"). When used in this way, the adjectives *obligate* and *facultative* are, by definition, mutually exclusive.

## TECHNIQUES FOR THE CULTURE OF ANAEROBIC BACTERIA

Techniques used for the cultivation of anaerobes depend upon the removal of free oxygen from the environment. The environment includes both the medium and the atmosphere above the medium. Strict obligate anaerobes require that virtually all the molecular oxygen be removed from the medium and from the atmosphere above the medium. Remember that oxygen diffuses easily and that any oxygen remaining in the atmosphere will rapidly dissolve in the medium and diffuse throughout. Strict anaerobiosis is difficult to obtain without special media and equipment.

There are many techniques that reliably remove most of the oxygen from the environment. There are a few techniques that reliably remove enough oxygen to allow growth of strict anaerobes. We will examine two of the techniques commonly used in clinical laboratories.

### Thioglycolate Medium

Thioglycolate medium is a nutrient medium that contains thioglycolic acid. This compound is a very strong reducing agent that reacts quickly with any oxygen that may diffuse into the medium. Of course, once the thioglycolic acid has been saturated with oxygen, the medium is no longer useful for the cultivation of anaerobes.

Thioglycolate medium is typically used as a semi-solid, tubed medium and is inoculated with a single stab into the center of the deep. It may, however, be used as a broth, or may be solidified with additional agar and used in plates. The use of a small amount of agar to make it semi-solid has the advantage of reducing the diffusion rate.

Some commercially available formulas for thioglycolate medium include an indicator dye that is colored in the presence of oxygen. The dye, typically methylene blue, appears blue when oxidized and is colorless when reduced. The use of such a dye has the obvious advantage of providing a convenient means of monitoring the anaerobiosis of the medium. However, there is evidence that the dye may inhibit the growth of some bacteria, especially certain gram-negative rods, and thioglycolate medium without indicator is used when these bacteria may be present.

When thioglycolate medium is correctly inoculated with a single stab down the center of the deep, the pattern of growth produced is distinct for each type of oxygen-related requirement (see Fig. E8-1a). Obligately aerobic bacteria will grow only in the aerobic zone (colored by dye, near the air/medium surface); obligate anaerobes will grow only in the anaerobic zone (uncolored by dye, in the bottom of the tube); and facultative bacteria grow along the entire length of the stab.

As noted earlier, oxygen dissolves readily in most media, including thioglycolate medium, and will quickly saturate the thioglycolic acid. Oxygen, being a gas, is also easily removed from

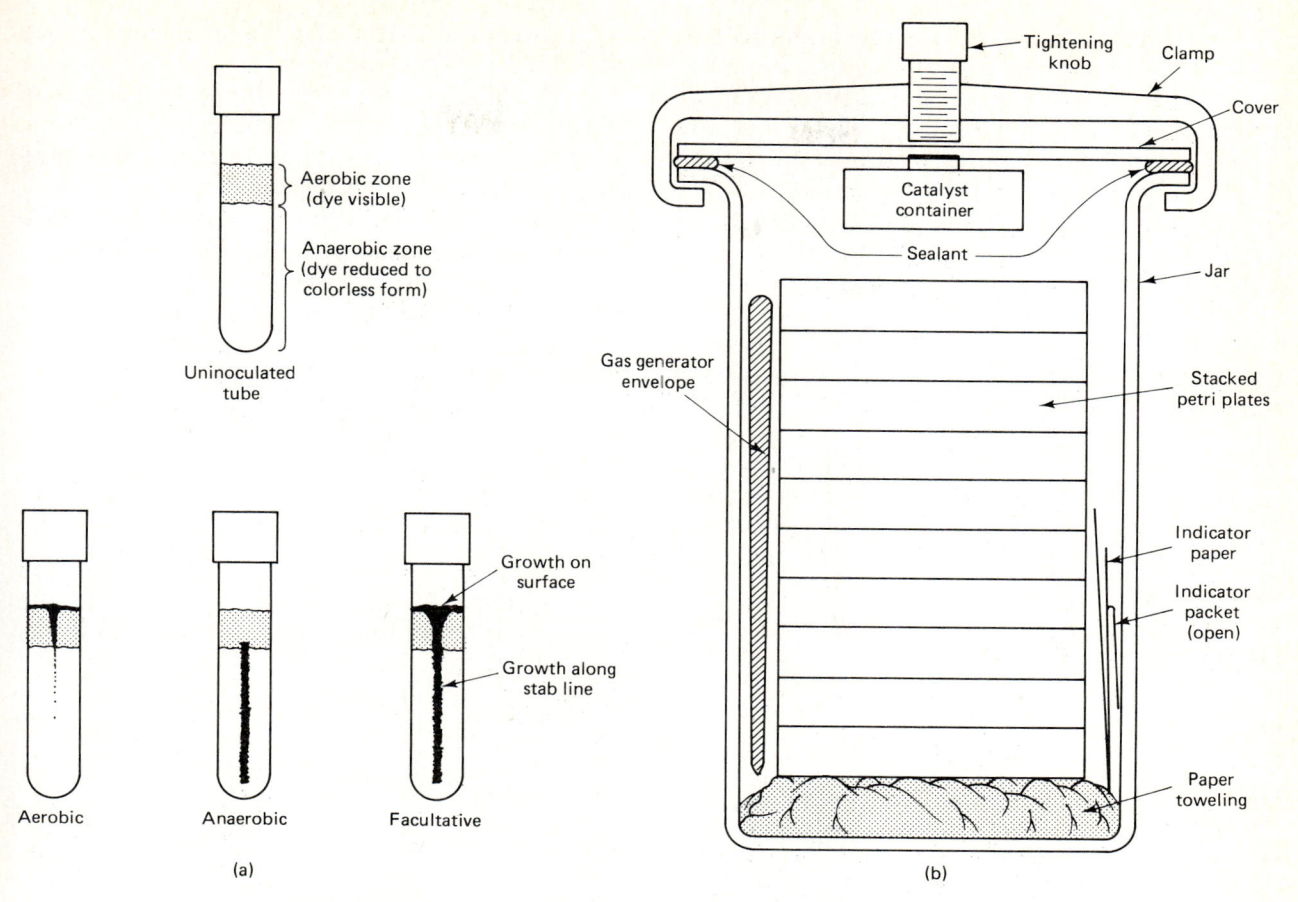

**Figure E8-1** (a) Growth in thioglycolate agar; (b) anaerobic jar (Gas-Pak).

solution by heating. If the dyed portion of the tube extends more than one quarter of the depth of the medium, the medium must be heated to boiling to remove the dissolved oxygen. The colored band will disappear upon heating, indicating that the oxygen has been removed. Indeed, many laboratories include this step whenever thioglycolate medium is used. It is, of course, essential if you use thioglycolate medium without indicator.

## Anaerobic Chambers or Jars

An anaerobic jar is a container that can be rendered anaerobic and which will remain anaerobic as long as it remains correctly sealed. Anaerobiosis is accomplished by using a compound, in this case hydrogen gas, that reacts with the oxygen inside the jar, usually reducing it to water. Once the atmospheric oxygen is depleted, the oxygen in the medium will rapidly diffuse into the atmosphere and, in turn, be reduced. We will use the Gas-Pak anaerobic system as an example. (Gas-Pak is a commercial brand name for a product of Baltimore Biological Laboratories, Inc.) The components of the Gas-Pak system (shown in Fig. E8-1b) are:

**Anaerobic jar.** The container consists of the jar itself, a cover, and a clamp that holds the cover tightly on the lip of the jar. Sometimes a sealant, such as petroleum jelly or stopcock grease, may be used to ensure a good, gas-tight seal between the jar and the cover.

**Catalyst.** The catalyst safely increases the rate of reaction between oxygen and the hydrogen gas mixture that is used to reduce it. The Gas-Pak system uses a catalyst that works safely at incubator temperature (in some older systems it was necessary to heat the catalyst electrically). It is usually contained in a small, screened container that is attached to the inner surface of the cover. The catalyst must be replaced occasionally.

**Gas generator.** The hydrogen-gas mixture that is used to reduce the oxygen is generated by chemicals provided in a foil envelope. It is only necessary to cut a corner off the envelope, add 10.0 ml of water, and place the envelope (with the water inside) into the Gas-Pak jar. The reaction that takes place in the jar produces the gas that will be used to reduce the oxygen. Each Gas-Pak envelope may be used only once.

**Indicator.** A disposable indicator is available for use in the Gas-Pak system. It consists of a piece of filter paper saturated with a methylene blue solution contained in a small foil packet. The methylene blue will turn colorless when the oxygen is removed from the atmosphere inside the jar. Although an indicator is not essential for the operation of the anaerobic system, it should be used to ensure that anaerobiosis was attained. (How else would you know if the catalyst was still effective?)

The Gas-Pak system operates in the following manner:

1. Obtain a Gas-Pak jar and ensure that the surfaces on the jar and cover that will mate when the jar is closed are clean and free of deep scratches and defects. Apply a very light coating of petroleum jelly or stopcock grease to one of the surfaces.
2. Check to ensure that a container of catalyst is attached to the inner surface of the cover. If one is not there, ask the instructor for a replacement.
3. Loosely pack one or two paper towels into the bottom of the anaerobic jar. (Moisture will be produced by the catalyzed reaction. The paper toweling will absorb most of the moisture.)
4. Stack the petri dishes into the jar, being careful that you do not try to use too many dishes. When closed, there should be a space of at least one inch between the top petri dish and the catalyst container.
5. Obtain a gas-generating envelope and cut the corner off along the dashed line (found in the upper corner of the envelope). Place the envelope in the jar, standing it up between the wall of the jar and the stacked petri dishes.
6. If you are using an indicator, peel back one side of the foil to expose the indicator paper and place it in the jar in such a way that you will be able to observe the color of the indicator paper through the wall of the jar. The paper will appear light blue. It should turn white when anaerobic conditions have been established.
7. Add 10.0 ml water to the gas-generator envelope. The tip of the pipette must not be forced into the envelope. You only need to insert the tip far enough to be sure that all the water flows into the envelope.
8. Quickly place the cover on the jar and clamp it in place. The tightening knob on the clamp should be lightly tightened, just enough to make a good seal between the jar and the cover. Do not overtighten the clamp.
9. Place the closed unit in the incubator and incubate for an appropriate period. (Some anaerobic cultures require up to 48 hours of incubation.) The Gas-Pak jar *must not be opened during the incubation period.*

## OBJECTIVES OF EXERCISE 7

The culture of anaerobic and microaerophilic bacteria requires specialized techniques for the removal of free oxygen. Many pathogenic bacteria are anaerobic or microaerophilic, and their isolation and identification is essential. To understand the nature of anaerobiosis and to appreciate the special techniques used, you should:

1. Understand the meaning of the terms *anaerobic, microaerophilic, facultative,* and *obligate* as used in this context.
2. Be familiar with simple anaerobic techniques, including thioglycolate medium and anaerobic jars.
3. Try to understand the significance of facultative versus obligate organisms in terms of the environmental conditions under which they can grow.

## MATERIALS NEEDED FOR THIS EXERCISE

1. Twenty-four- to 48-hour cultures of the following organisms:
   *Clostridium sporogenes*
   *Bacillus subtilis*

*Pseudomonas aeruginosa*
*Staphylococcus aureus*
*Escherichia coli*
*Micrococcus luteus*

2. Seven screw-capped tubes (deeps) of thioglycolate medium. If the dye is visible for more than one quarter of the depth of the agar deep, the tubes will need to be heated in a boiling-water bath to drive off the dissolved oxygen. After you heat the tubes, do not shake them or disturb them while they are cooling.
3. Sufficient plates of nutrient agar or other nutrient medium to make two streak plates of each of the organisms used in this exercise. You should make one set of streak plates for anaerobic and one for aerobic incubation. You might want to reduce the number of plates used (which you must fit into the anaerobic jars) by streaking two organisms on each plate.
4. Gas-Pak anaerobic jars, with catalyst, indicator, and gas-generator envelopes.
5. 10.0-ml pipettes.

## LABORATORY PROCEDURE

1. Obtain one tube of thioglycolate medium for each organism to be tested, plus one tube for use as a control.
2. Label all the tubes and mark the depth of the dye on the control tube with a wax pencil or a felt-tip pen.
3. Inoculate all tubes (except, of course, the control) with the test organisms by making a single stab down the center of the deep. Use an inoculating needle or a sterile Pasteur pipette.
4. Make two full (or half, as instructed) streak plates of each organism being tested. Place one set of the petri plates in the anaerobic jar. The other set will be incubated under aerobic conditions.
5. Set up one Gas-Pak anaerobic jar according to the instructions given in this exercise. Details will be given in the Laboratory Protocol section later in the exercise.
6. Incubate all cultures for 24 to 48 hours. It is important that the anaerobic jar remain sealed for the entire incubation period. (Why can't you open the jar to check your cultures?)
7. Record the growth pattern observed in the thioglycolate medium. Direct your attention to position of any visible growth relative to the appearance of dye. Measure the change in the depth of the aerobic zone.
8. Examine the petri-plate cultures, noting colonial morphology of those cultures that did grow. Complete gram stains as appropriate. Was there any correlation between the growth observed in thioglycolate medium and in the anaerobic jar? Compare the growth and colonial morphology of the aerobic and anaerobic cultures.
9. Record your results on the Laboratory Report Form for this exercise.

## REFERENCES

### Text References

1. Atlas, Ronald M., *Microbiology: Fundamentals and Applications.* Chapters 5, 10, and 16.
2. Brock, Thomas D., David W. Smith, and Michael T. Madigan, *Biology of Microorganisms,* 4th ed. Chapters 4, 8, 13, 14, and 15.
3. Jensen, Marcus, *Introduction to Medical Microbiology.* Chapters 6 and 20.
4. Nester, Eugene W., et al., *Microbiology,* 3rd ed. Chapters 6, 13, and 19.
5. Stanier, R. Y., E. A. Adelberg, and J. Ingraham, *The Microbial World,* 4th ed. Chapter 10.
6. Stanier, R. Y., et al., *Introduction to the Microbial World.* Chapter 5.
7. Wistreich, George A., and Max D. Lechtman, *Microbiology,* 4th ed. Chapters 6, 7, 27, 33, and 34.

### Resource References

1. Finegold, Sidney M., and W. J. Martin, *Bailey and Scott's Diagnostic Microbiology,* 6th ed. Chapters 13 and 27–30.
2. Gerhardt, P., ed., *Manual of Methods for General Bacteriology.* Chapter 6.
3. Lennette, E. H., ed., *Manual of Clinical Microbiology,* 3rd ed. Chapters 35 through 40.
4. McGonagle, L. A., *Procedures for Diagnostic Microbiology.* Consult specific topical references.

## LABORATORY PROTOCOL

### Exercise 8 — Growth of Anaerobic Bacteria

Check each step when you complete it.

**First Day**

_____  1. Obtain and label one screw-cap tube of thioglycolate medium for each organism used in this exercise. Also obtain a control tube and mark the depth of the visible dye with a wax pencil.

*Note:* If the dye is visible for more than one-half inch into the medium, the dissolved oxygen must be driven off by heating the tubes in a boiling-water bath.

_____  2. Inoculate the tubes, using a needle or sterile Pasteur pipette, with a single stab into the center of the deep. The following 24- to 48-hour cultures may be used:

*Clostridium sporogenes*
*Bacillus subtilis*
*Pseudomonas aeruginosa*
*Escherichia coli*
*Staphylococcus aureus*
*Micrococcus luteus*

_____  3. Tightly close the caps on the tubes, then incubate the cultures for 24 to 48 hours.

_____  4. Set up one Gas-Pak anaerobic jar:

_____  a. Obtain all the component parts (jar, cover, catalyst, indicator, and gas generator).

_____  b. Examine the jar, being careful to ensure that the surface that mates with the cover is clean and free of defects. Wipe a small amount of petroleum jelly or stopcock grease on the mating surfaces.

_____  c. Examine the cover. Make sure that the catalyst container is properly attached and that it contains catalyst pellets. If you have any doubts, ask your instructor.

_____  d. Place a few (2-3) pieces of paper toweling in the bottom of the jar (to absorb moisture).

_____  5. Make two streak plates of each of the organisms used in this exercise. If you streak two organisms on each plate, consider using the following combinations (why?):

*M. lutea* and *S. aureus*
*P. aeruginosa* and *E. coli*
*B. subtilis* and *C. sporogenes*

_____  6. Place one set of the petri plates in the anaerobic jar and proceed as follows:

_____  a. Cut the corner off a gas-generator envelope and position it between the petri plates and the wall of the anaerobic jar.

_____  b. Peel back one side of the foil packet containing indicator paper. Position it so the paper can be observed from the outside of the jar.

_____  c. Using a pipette, place 10.0 ml of distilled water into the gas-generator envelope. Do not allow the tip of the pipette to tear or otherwise damage the envelope.

92   EXERCISE 8

    _____ d. Quickly close the jar and clamp the cover in position. Do not overtighten.

_____ 7. Place the anaerobic jar in the incubator and incubate for 24 to 48 hours. Also incubate the aerobic set of plates for 24 to 48 hours.

*Note:* As the catalyst works, the cover of the jar will get noticeably warm and moisture will condense on the inner surfaces of the jar. This indicates that the catalyst is working and is completely normal. The indicator paper should turn from light blue to white in about three hours.

**Second Day**

_____ 8. Record the depth of the dye in the thioglycolate medium. Calculate how far the oxygen diffused in 24 and 48 hours.

_____ 9. Observe the pattern of growth in thioglycolate medium. Record whether growth occurred in the aerobic part of the tube, the anaerobic part, or both parts (see Fig. E8-1a).

_____ 10. Before opening the anaerobic jar, note the color of the indicator paper. It should be white; if it is not, you must assume that anaerobiosis was not maintained.

_____ 11. Study and record the colonial morphology of all colonies. Try to correlate these results with the observations from the thioglycolate medium. Confirming smears and gram stains should be made on all cultures.

_____ 12. Compare the growth observed on the two sets of petri plates. Describe any differences in morphology or growth.

_____ 13. Record your results on the Laboratory Report Form and answer all the questions in the report.

# NOTES

Name: _____

## LABORATORY REPORT FORM

### Exercise 8—Growth of Anaerobic Bacteria

1. Complete the following table:

    **GROWTH OF SELECTED BACTERIAL SPECIES UNDER ANAEROBIC CONDITIONS**

    | Organism | Thioglycolate (Position of Growth) | Anaerobic Jar (Indicate + or −) |
    |---|---|---|
    | S. aureus | _____ | _____ |
    | M. luteus | _____ | _____ |
    | B. subtilis | _____ | _____ |
    | C. sporogenes | _____ | _____ |
    | P. aeruginosa | _____ | _____ |
    | E. coli | _____ | _____ |

2. Describe any differences in colonial morphology observed between the aerobic and anaerobic cultures.

3. Using the definitions provided in this exercise, indicate which of the following bacteria are obligate aerobes, obligate anaerobes, or facultative.

    a. *Staphylococcus aureus*

    b. *Micrococcus luteus*

    c. *Bacillus subtilis*

    d. *Clostridium sporogenes*

e. *Pseudomonas aeruginosa*

f. *Escherichia coli*

**ANSWER THE FOLLOWING QUESTIONS**

1. List four pathogenic species of *Clostridium* and the disease that each causes.
    a.

    b.

    c.

    d.

2. List four other pathogenic bacteria that are anaerobic. (Consult the References.)
    a.

    b.

    c.

    d.

3. Explain why an organism might be an obligate anaerobe.

4. Explain why an organism might be an obligate aerobe.




5. List two energy-producing pathways that do not require oxygen.

# Exercise 9
# Development and Use of Diagnostic Keys

A diagnostic key will help you to determine the genus and species of an organism by comparing the characteristics of one organism (the unknown) with those of organisms used as reference bacteria (type cultures). The purpose of diagnostic keys is to match your unknown with organisms having the same characteristics. You must, in effect, compare your isolate with others and determine which of the others it most closely matches.

The use of phenotypic characteristics to match the phenotype of one organism with the phenotype of the reference organism is an application of the concepts of *Adansonian* or *numerical taxonomy.* This concept assumes that each phenotypic trait is of equal value in determining taxonomic identity and that the best "fit" is obtained when the largest number of traits are compared. Of course, once the characteristics of the reference organism are known, they can be recorded and the unknowns identified by using written descriptions and keys.

The list of characteristics you will use for your data-base exercise was obtained in the previous exercises and includes most, if not all, of the following:

    Cell morphology
    Staining reactions
    Colony morphology
    Biochemical characteristics

Two common applications of Adansonian taxonomy in modern bacteriology are the *dichotomous key* and *profile analysis*.

- The *dichotomous key* is a list of increasingly specific phenotypic traits that will allow you to assign a genus and species name to your organism by a stepwise elimination of other organisms. In this system, the identifying characteristics are used one at a time to eliminate unrelated organisms.
- Phenotypic *profile analysis* considers several phenotypic traits simultaneously and matches the phenotypic profile you have developed with that of other organisms whose profiles are known. In this system, as many identifying characteristics as is practical are used simultaneously.

## THE DICHOTOMOUS KEY

The dichotomous key is, in effect, a road map that takes you to the genus and species name of your unknown. It requires that you be able to define your isolate by answering simple, single-variable questions about it. Usually, a dichotomous key's questions can be answered with a simple yes or no. For example, "Does the unknown ferment lactose?" or "Does the unknown hydrolyze gelatin?" In other cases, the question calls for a choice between two mutually exclusive conditions or states, such as "Is it gram-positive or gram-negative?" or "Is it a rod or a coccus?" When you put a dichotomous key into diagram form, it should look like a pyramid, the first test being at the top or apex, and the last tests and the names of the identified bacteria at the base.

In order to use a dichotomous key, you need only know which tests must be performed and then follow the key to determine the correct genus and species. Obviously, however, someone must have first determined which tests could be applied diagnostically and then determined the best sequence in which to use the tests. As you examine the sample key, note that the tests applied late in the key are much more specific than those used early in the key. The later tests usually distinguish single species, while the early ones typically distinguish genera or even groups of genera.

The disadvantages of dichotomous keys include the necessity to interpret the tests absolutely as either plus/minus (+/−) or yes/no. There is no accommodation of variant organisms, or those with atypical characteristics differing from the type strain. It is also true that a single error in test interpretation can result in misdiagnosis. Each dichotomous key is based on judgments about the proper sequence of tests used in the key. In a very real way, the dichotomous key is like a road map—there may be several routes to the same destination, but a wrong turn early in the route will leave you lost in a strange neighborhood.

## PHENOTYPIC PROFILE ANALYSIS

Profile analysis is a diagnostic procedure that simultaneously compares several tests to develop a profile of the unknown and compare it with profiles in a reference data bank.

The advantages of profile analysis include the ability to compare many tests at once, avoiding the problems created by "wrong turns," and the ability to accommodate variants that have one or more atypical characteristics. The disadvantage of this technique is that as additional characteristics are added to the profile, it is increasingly difficult to keep everything straight. Although the average human mind can reasonably handle up to ten or so traits, as many as 20 or 30 individual traits must often be used to complete a diagnosis. The development of computers has made profile analysis readily available; several commercial systems are on the market. One such system, produced by Analytab Products, Inc., uses 22 tests to determine a seven-digit code for identification of the *Enterobacteriaceae*. The numerical code is compared against similar numerical profiles for over 27,000 reference strains, and the most probable match is selected.

## AN EXAMPLE EXERCISE: DEVELOPMENT OF A DICHOTOMOUS KEY

You will, as noted above, use the characteristics noted in the previous four exercises to assemble a *bacterial characteristic data matrix*. That data will then be used to write your own dichotomous key. In the example given here, all the necessary information is provided. You should try to understand that the three charts are simply three ways of presenting the same data.

You should begin by developing a data matrix containing all the information you have accumulated about each organism. List the organism names on the vertical axis and the tests and other data points on the horizontal axis as shown in the following example.

### DATA MATRIX CHART—AN EXAMPLE

| Name of Organism | Gram Reaction | Cell Shape | Ferments Lactose | Liquefies Gelatin | Coagulase Test |
|---|---|---|---|---|---|
| E. coli | − | Rod | + | − | − |
| S. aureus | + | Coccus | − | + | + |
| S. epidermidis | + | Coccus | − | − | − |
| P. aeruginosa | − | Rod | − | − | − |

Your matrix will, of course, be much more extensive than the example shown here, since you will use all the data (there are 29 characteristics listed on the Bacterial Characteristics Form) accumulated in the previous exercises. After the matrix is completed, you can begin to develop your dichotomous key.

For the organisms shown in the example, the first branch in the key might be the gram reaction and/or the shape of the cells. The rods can then be differentiated on the basis of the fermentation of lactose, and the cocci can be distinguished by either the coagulase reaction or the liquefication of gelatin.

The information contained in the Data Matrix Chart is presented below as it might appear in a flow diagram or as a "road map." Try to determine the relationship between the two ways of presenting these data.

The first branch in your key should be a characteristic that divides the group of organisms into two large, approximately equal groups. At each subsequent branch of your key, you should use increasingly more specific tests until individual organisms are identified. When you have completed the key, you should be able to use it to identify any of the bacteria that you worked with in the previous exercises.

**DICHOTOMOUS KEY—AN EXAMPLE**

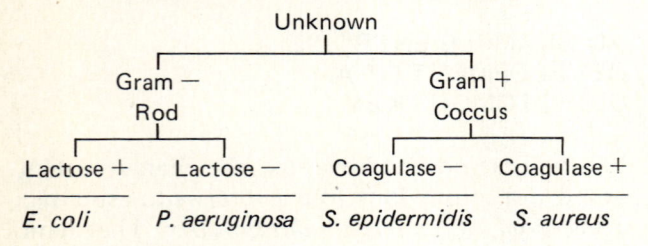

```
                    Unknown
            ┌──────────┴──────────┐
          Gram −                Gram +
           Rod                  Coccus
        ┌────┴────┐          ┌────┴────┐
    Lactose +  Lactose −   Coagulase − Coagulase +
    E. coli   P. aeruginosa S. epidermidis S. aureus
```

There is yet another way of presenting the dichotomous key. It takes the form of a series of statements about the organism; it is probably the form that you will most likely encounter. The following table is a short dichotomous key using the examples in this exercise. Other examples of dichotomous keys are given in Exercises 11, 12, and 13.

**THE DICHOTOMOUS KEY**

1. Organism is a gram-negative rod
    a. Ferments lactose — E. coli
    b. Does not ferment lactose — P. aeruginosa
2. Organism is a gram-positive coccus
    a. Organism is coagulase positive — S. aureus
    b. Organism is coagulase negative — S. epidermidis

## PHENOTYPIC PROFILE ANALYSIS— A BRIEF DISCUSSION

If we use the example given above, a profile-analysis system would consider all five of the characteristics simultaneously. An organism that gave the results shown on the chart (the unknown) would be matched with *S. aureus*.

**DATA MATRIX CHART—AN EXAMPLE**

| Name of Organism | Gram Reaction | Cell Shape | Ferments Lactose | Liquefies Gelatin | Coagulase Test |
|---|---|---|---|---|---|
| E. coli | − | Rod | + | − | − |
| S. aureus | + | Coccus | − | + | + |
| S. epidermidis | + | Coccus | − | − | − |
| P. aeruginosa | − | Rod | − | − | − |
| Unknown | + | Coccus | − | + | + |

If one test varied from the above reference profile, the system would report the "goodness of fit" (e.g., "four/five" match) with *S. aureus*. In this case, however, since there are only two tests differentiating *S. aureus* from *S. epidermidis*, the profile analysis system would alert you if the atypical test were one of the two needed to differentiate between these two bacteria.

## LABORATORY OBJECTIVES

The requirements for this exercise are such that you will need to become familiar with the use of phenotypic characteristics as tools for bacterial identification. To accomplish this, you will:

1. Develop a matrix chart for all the phenotypic characteristics of the bacteria studied in the previous exercises.
2. Use these phenotypic characteristics to develop a diagnostic key that would allow anyone to identify any of the bacteria studied so far.
3. Gain an understanding of how the identification of unknown bacterial isolates must follow a logical, step-by-step process.

4. Gain an understanding of the differences, both advantageous and disadvantageous, between a dichotomous key and a phenotypic profile.

## LABORATORY PROCEDURE

You should complete the following assignments:

1. Complete a data matrix chart for all the bacterial species studied in the previous exercises. Include every possible diagnostic tool, from shape of the cell to individual enzymatic reactions. Use the data from your Bacterial Characteristics Forms to construct the charts.
2. Develop your own dichotomous key from this data.
3. Try using the key on two or three of your known organisms to see if it works. (Better yet, have another student try it.)
4. Make sure there is no ambiguity in the key. If a test is used more than once, consider if it can be moved up to a branch closer to the apex.

## REFERENCES

### Text References

1. Atlas, Ronald M., *Microbiology: Fundamentals and Applications.* Chapter 11. Appendix III.
2. Brock, Thomas D., David W. Smith, and Michael T. Madigan, *Biology of Microorganisms,* 4th ed. Chapter 18. Appendix 4.
3. Jensen, Marcus, *Introduction to Medical Microbiology.* Chapter 6.
4. Nester, Eugene W., et al., *Microbiology,* 3rd ed. Chapter 11. Appendixes IV and V.
5. Stanier, R. Y., E. A. Adelberg, and J. Ingraham, *The Microbial World,* 4th ed. Chapter 16.
6. Stanier, R. Y., et al., *Introduction to the Microbial World.* Chapter 9.
7. Wistreich, George, and Max D. Lechtman, *Microbiology,* 4th ed. Appendix A.

### Resource References

1. Buchanan, R. E., and N. E. Gibbons, eds., *Bergey's Manual of Determinative Bacteriology,* 8th ed.
2. Finegold, Sidney M., and W. J. Martin, *Bailey and Scott's Diagnostic Microbiology,* 6th ed. Chapter 5 (consult specific references).
3. Gerhardt, P., ed., *Manual of Methods for General Bacteriology.* Chapters 20 through 22.
4. Lennette, E. H., ed., *Manual of Clinical Microbiology,* 3rd ed. Chapter 1.

# LABORATORY PROTOCOL

## Exercise 9—Development and Use of Diagnostic Keys

Check each step when you complete it.

_____ 1. Prepare a data matrix chart for all the bacterial species examined in the previous exercises. The tests are usually written across the top, with the names of the bacteria written down the left side of the chart.

_____ 2. Write a diagnostic key to identify the organisms on your list, using the one shown on page 100 as a model.

_____ 3. Test your key (or better yet, have one of your classmates test it) against one or more of the organisms listed (assume it to be an unknown).

# NOTES

Name: _____

# LABORATORY REPORT FORM

## Exercise 9—Development and Use of Diagnostic Keys

1. There are no prepared forms for this exercise. Your approach to the development of a diagnostic key will be sufficiently individualized to make preparation of a single form impossible.
    Study the examples in the exercise and try to develop your own. Remember that, as with any road map, there is more than one way to get to any given destination.
2. You might consider simplifying your task by starting with certain assumptions:
    a. Can a separate key be used for major groups of bacteria, such as the gram-positive cocci, the gram-negative rods, and so on?
    b. Determine which reactions divide the largest number of organisms into manageable groups. For example, lactose fermentation, oxidase reaction, and hydrogen sulfide production might be used to separate the gram-negative rods into manageable groups, especially if the reactions are used sequentially.
    c. Use the approach that you are solving an unknown (as you will soon be doing). What is the most direct route to the name of the unknown?

# Exercise 10
# Use of Diagnostic Keys: Bacterial Unknowns

One of the reasons you need to be familiar with the development of diagnostic keys is that you may have to apply that information to the solving of unknowns. In clinical microbiology, every specimen sent to the laboratory for analysis is an unknown. It is often necessary to identify nonclinical isolates, because pathogenic bacteria are frequently found in environmental samples and in contaminated food and water. In short, when you send a culture from any source to the laboratory, asking them to identify the predominant organism(s), you have presented them with an unknown.

## PROBLEM-SOLVING APPROACH TO UNKNOWNS

The first procedures used in any identification protocol must be those that isolate the organisms into pure culture. The technician must be familiar with the bacteria likely to be encountered and be able to recognize those bacteria by their colonial morphology. This initial isolation is sometimes referred to as primary isolation.

After isolation is achieved, the identification of the unknown proceeds by an orderly and stepwise elimination of alternate choices. The first choices are on the basis of the cellular morphology and staining reactions of the isolated culture. At the very least, a gram stain is considered essential, followed by whichever other stains would prove helpful (e.g., acid-fast and metachromatic-granule stains).

Once the cell morphology, the staining reactions, and the colonial morphology are known, an educated guess can be made as to the possible genus and species of the bacteria. Then, depending on the possibilities suggested by the information already available, specific biochemical tests are used to confirm the identity of the unknown or to determine the genus and species if it is not already known. Which biochemical tests do you use? That's where your knowledge of and experience with diagnostic keys will prove useful. A good diagnostic protocol will use the least number of tests to arrive at the solution. To do this, you must know which tests are important for the type of organisms you are trying to identify. For example, you would use a different set of tests for gram-positive cocci than you would use for gram-negative rods.

### PROCEDURE FOR SOLVING UNKNOWNS

*First Step*

Primary isolation: Streak plates for initial isolation and determination of colonial morphology

*Second Step*

Staining reactions: Microscopic examination for determination of cellular morphology and staining reactions

*Third Step*

Biochemical tests: Confirmation and identification by biochemical characterization, using a carefully determined set of tests

The diagnostic key that you developed in the previous exercise may be used to identify your unknowns or to quickly determine if the unknown you are working with is not one of the species included in the key. You may need to consult the Resource References and use the keys contained therein. The Resource References listed in Appendix 1 contain several diagnostic keys that may prove useful.

The preferred way to solve an unknown is to carefully determine which tests, and in what order, should be used. The worst way, and often the most confusing way, to handle an unknown is to simply subject it to every available test, hoping that sufficient data will be generated to provide the correct answer. It is true that all of the needed data will have been generated, but because there was no careful stepwise elimination of alternative choices, the student will usually be confronted with what often seems to be hopeless confusion.

In clinical applications, the identification of the bacteria in submitted samples is usually expected within 72 hours, and often within 48 hours. Speed and accuracy are essential and can only be maintained if a planned protocol is followed. If you are working with a gram-negative rod, you do not use tests that are useful for the identification of gram-positive cocci. If a gram-positive rod that looks like a diphtheroid is observed to produce metachromatic granules, you do not need to do the oxidase test, but you do need to use a few selected tests to confirm the identification of *Corynebacterium diphtheria*. On the other hand, if you isolated a gram-positive coccus, you would consider using the coagulase test, catalase test, mannitol fermentation, and gelatin liquefaction (in that order?).

In summary, the best approach to identifying a bacterial unknown is a problem-solving approach. After each test you must ask yourself "What possibilities are indicated by the results?" and "What tests are needed to further differentiate this isolate?" The surest way to become completely lost and confused is to attempt to apply tests in a shotgun fashion, hoping to hit upon the answer. Plan your approach carefully.

## LABORATORY OBJECTIVES

You will be asked to identify unknown bacterial cultures. To successfully accomplish that task, you must:

1. Apply all of the skills learned in the previous five exercises, including isolation of pure cultures, microscopic examination of stained smears, and recognition of colonial morphologies.
2. Use a problem-solving approach to develop a protocol for the orderly application of diagnostic tests and the interpretation of the results obtained from those tests.
3. Learn to anticipate what's next when the results of one set of tests are known.

## MATERIALS NEEDED FOR THIS LABORATORY

*Note:* The bacterial cultures should include, but not necessarily be limited to, bacteria covered in the previous exercises.

1. Unknown cultures: The cultures should be between 24 and 48 hours old. If they are mixed, the mixing should be done just before the unknowns are distributed. (Some bacteria inhibit growth and/or produce substances that are toxic to other bacteria.) Your unknowns should include:
    a. Pure cultures: at least one species, and
    b. Mixed cultures: at least one, containing three bacterial species.
2. Nutrient agar plates.
3. Staining reagents for Gram's, acid-fast, and metachromatic-granule stains.
4. Media and chemical reagents as needed to identify the unknowns.

## LABORATORY PROCEDURE

1. The protocol you will follow for this laboratory exercise will be one you develop yourself. *You* must decide upon the protocol as you acquire data about your unknowns. *You* must decide what to do next.
2. The following initial steps are recommended:
    a. Make gram stains and streak plates of all cultures.
    b. On the basis of the gram stains, and while you are waiting for the streak plates to grow, plan what tests will be used next. Consult your charts and keys from Exercise 9.

3. Continue with the testing until you have identified each organism in both the pure cultures and the mixtures.
4. Complete a Bacterial Characteristics Form for each organism, filling in only those results used in the diagnosis.

## ADDITIONAL HINTS AND SUGGESTIONS

1. *Never* discard a culture until any subcultures of it have grown up.
2. *Always* make a streak plate of your isolated cultures (even if they were derived from a single, well-isolated colony).
3. *Always* keep a culture of your unknown isolates (a stab slant is preferred) until you are completely finished with them. You may need to refer back to them.
4. *Always* keep all slides of your unknowns until you are completely finished with them.
5. *Always* include controls in your tests. For example, use uninoculated tubes of the fermentation media to compare colors after incubation. Use an uninoculated tube of gelatin as a negative control.
6. *Always* keep detailed notes of your laboratory work. It is almost always necessary to go back over them to determine some previous test results or to see if the results were interpreted correctly.
7. *Always keep calm.*

## REFERENCES

### Text References

1. Atlas, Ronald M., *Microbiology: Fundamentals and Applications.* Chapters 11 and 16. Appendix III.
2. Brock, Thomas D., David W. Smith, and Michael T. Madigan, *Biology of Microorganisms,* 4th ed. Chapter 18. Appendix 4.
3. Nester, Eugene W., et al., *Microbiology,* 3rd ed. Chapter 11. Appendixes IV and V.
4. Wistreich, George A., and Max D. Lechtman, *Microbiology,* 4th ed. Appendix A.

### Resource References

Use as general resource references for specific organisms and/or for specific test procedures:

1. Buchanan, R. E., and N. E. Gibbons, eds., *Bergey's Manual of Determinative Bacteriology,* 8th ed.
2. Finegold, Sidney M., and W. J. Martin, *Bailey and Scott's Diagnostic Microbiology,* 6th ed.
3. Gerhardt, P., ed., *Manual of Methods for General Bacteriology.*
4. Lennette, E. H., ed., *Manual of Clinical Microbiology,* 3rd ed.
5. McGonagle, L. A., *Procedures for Diagnostic Microbiology.*

## LABORATORY PROTOCOL

### Exercise 10—Use of Diagnostic Keys: Bacterial Unknowns

Check each step when you complete it.

**First Day**

_____ 1. Make gram stains of all cultures.

_____ 2. Make streak plates of all cultures. Incubate for 18–24 hours.

_____ 3. Use the information from the gram stain to plan your subsequent steps.

**Second and Subsequent Days**

_____ 4. Complete identification of the unknown(s).

_____ 5. Complete the Laboratory Report Form.

# NOTES

Name: _____

## LABORATORY REPORT FORM

### Exercise 10—Use of Diagnostic Keys: Bacterial Unknowns

1. Complete the identification of your unknown(s).
2. Keep detailed notes of each step.
3. Write a step-by-step protocol for each of your unknowns, including isolation and staining procedures used.
4. Complete one Bacterial Characteristics Form for each organism, but use only those tests that were needed for its identification.

# Exercise 11
# Throat Cultures

## ISOLATION OF PATHOGENIC GRAM-POSITIVE AND GRAM-NEGATIVE COCCI AND DIPHTHEROIDS

The upper respiratory tract has a rich population of indigenous organisms, most of which are considered to be normal flora. It also supports the growth of several important pathogens, the most important of which include the streptococci, the staphylococci, the neisseriae, the diphtheroids, yeasts (particularly *Candida*), and enteric gram-negative rods. In addition to these, which are commonly isolated from the upper respiratory tract of persons of all ages, certain other pathogens are occasionally encountered in samples from the very young (under five years) and the elderly (over sixty years). These latter include the genera *Hemophilus, Bordatella, Yersinia,* and *Francisella*.

In general, techniques for the isolation and identification of the pathogenic forms have been well defined, and the procedures are straightforward and direct. Much of the diagnostic work involves distinguishing between the pathogenic and nonpathogenic species that comprise the normal flora. The use of special differential and selective media has greatly facilitated this task.

Several of these media will be used in this exercise. You will be asked to take throat cultures from your partner and to compare the growth obtained from them with the characteristic colonies produced by known and representative organisms. The selective and differential media you will use in this exercise include:

**1. Blood agar.** Blood agar is a nutrient medium, such as trypticase soy agar, that has whole sheep red blood cells added to it. The most common mixture uses a 5% suspension of specially washed cells in the medium. Of course, special preparatory procedures must be followed to prevent damage to the red blood cells by the heat of the melted medium. It is not uncommon to add special nutrients and/or selective agents to blood agar, depending upon the needs of the laboratory.

**2. Chocolate agar.** Chocolate agar is blood agar that has been heated to denature the proteins of the sheep red blood cells. This causes them to lyse and to turn a light brown color, not unlike that of milk chocolate (hence the name). Most formulas for chocolate agar include certain enrichment factors (such as added hemoglobin and other factors) that encourage the growth of some of the more fastidious pathogens, including *Neisseria* and *Hemophilus*.

**3. Thayer-Martin or Martin-Lewis media.** These two media are chocolate agar that has had certain antibiotics (vancomycin, colistin, and mystatin or anisomycin) added to it to make it very selective for *N. gonorrhoeae* and *N. meningitidis*. These media are so selective that growth of gram-negative cocci on either of these is considered to be presumptive evidence of the presence of these pathogens.

**4. Mannitol-salt agar.** Mannitol-salt agar is a medium containing the sugar mannitol and 7.5% sodium chloride (salt). The salt inhibits the growth of most bacteria *except* the staphylococci. The mannitol differentiates between *S. aureus* (ferments mannitol) and *S. epidermidis* (does not ferment mannitol).

## THROAT CULTURES: SITE TO BE SAMPLED

Throat cultures are taken from the area at the very rear of the mouth, behind the uvula (often falsely referred to as the tonsil) by rotating a sterile swab over the area. Any inflamed area that might be observed should be cultured. It is very important that:

1. The culture be taken from the correct part of the oral cavity. The single most common error encountered in throat cultures is that the culture is taken from the wrong part of the mouth. The swab should be *rotated* over the mucosal surface *behind* the uvula.
2. The swab must be rotated both when the sample is taken and when the swab is used to inoculate the media. This will ensure that a large enough surface area will be sampled and that the media will be adequately inoculated.

## TECHNIQUES FOR MEDIA INOCULATIONS

The correct method for inoculating diagnostic media when the sample is contained on a swab is to rotate the swab on the surface of the medium so that the entire surface of the swab has had contact with the medium. You should cover an area about the size of a nickel or quarter and then use an inoculating needle or loop to complete a streak plate. Use the four-quadrant streak plate discussed in Exercise 1.

The following selective and differential media are commonly used for routine throat cultures:

1. Blood agar
2. Chocolate agar
3. Martin-Lewis or Thayer-Martin agar
4. Mannitol-salt agar

## REACTIONS ON SELECTIVE AND DIFFERENTIAL MEDIA

**1. Blood agar.** Bacteria that secrete hemolytic enzymes are able to lyse the sheep red blood cells in the medium. The destruction of these cells will produce distinct zones of hemolysis around their colonies. They may be categorized according to the type of hemolysis they cause:

- Alpha-hemolysis: an incomplete lysis of the erythrocyte and a partial denaturation of the hemoglobin to various heme products. Alpha-hemolysis results in green discoloration of the area around the colony. Microscopic examination would reveal cellular debris in the green-colored regions.
- Beta-hemolysis: a complete lysis of the erythrocyte and complete denaturation of the hemoglobin to colorless products. Beta-hemolysis results in a completely cleared zone around the colony, and no cellular debris would be observed upon microscopic examination.
- Nonhemolytic: There is no visible change in the erythrocytes around the colonies. Sometimes nonhemolytic streptococci are referred to as *gamma-hemolytic,* but since there is no change in the erythrocytes, *nonhemolytic* is a more accurate description.

**2. Chocolate agar and Martin-Lewis/Thayer-Martin media.** Some bacterial hemolysins that are able to denature hemoglobin in blood agar are also able to cause changes in the color of the chocolate agars. Many streptococci and lactobacilli that cause either alpha- or beta-hemolysis will cause either a greening or a decoloration of the light brown color in chocolate agar. However, if the bacteria being isolated are nonhemolytic, there will be no color change in the medium, and their colonial morphology will be typical of whatever species is being isolated.

**3. Mannitol-salt agar.** Mannitol-salt agar is selective for the staphylococci, although any isolate should be confirmed with a gram stain and the catalase test (see below). It is also differential in that it will distinguish between *S. aureus* and *S. epidermidis* on the basis of the fermentation of mannitol. Organisms that ferment mannitol will produce a yellow zone around the colony as a result of the production of acids during fermentation. Organisms that do not ferment mannitol will produce a deeper red color (alkaline reaction) or will not change the color at all.

## DIFFERENTIATION AND IDENTIFICATION: SCREENING AND CONFIRMING TESTS

**Streptococci from other gram-positive cocci.** All members of the genus *Streptococcus* are catalase negative, and all other gram-positive cocci are catalase positive. The catalase test (Exercise 7) is performed by placing one drop of hydrogen peroxide on a colony. Catalase produced by positive organisms will cause the hydrogen peroxide to be broken down to water and oxygen, resulting in vigorous bubbling. Catalase-negative organisms do not cause bubbling.

**Staphylococci from micrococci.** All of the staphylococci are facultative, while most of the common micrococci are obligate aerobes. The two genera can be differentiated by determining their growth characteristics in semi-solid thioglycolate deeps (see Exercise 8). The staphylococci will grow throughout the entire length of the stab, while the micrococci will only grow at the top.

**Staphylococcus aureus from Staphylococcus epidermidis.** Any one of the following tests can be used to differentiate between *S. aureus* and *S. epidermidis*. The coagulase test, described in Exercise 7, is considered the most important, and it is the one most frequently used.

| Test | S. aureus | S. epidermidis |
|---|---|---|
| Coagulase | + | − |
| Acid from mannitol | + | − |
| Liquefies gelatin | + | − |
| DNAase positive | + | − |

**Alpha-hemolytic streptococci.** Many of the alpha-hemolytic streptococci are part of the normal flora. However, many of them can be pathogenic when they are introduced into other parts of the body; one of them, *Streptococcus pneumoniae*, is the causative agent of bacterial pneumonia and can be a very serious pathogen. It must always be ruled out when alpha-hemolytic streptococci comprise a significant proportion of the isolates from a throat culture. Virtually all the alpha-streptococci except *Streptococcus pneumoniae* are resistant to lysis by bile-salt solutions and will grow in the presence of such salts. This characteristic is used to determine if any of the alpha-hemolytic streptococci that are cultured from the throat are *Streptococcus pneumoniae*. The sensitivity of the bacterium to bile salts can be measured by either placing one drop of bile-salt solution over a colony or by using prepared disks (Taxo P or Optochin). Sensitive organisms will dissolve in the bile solution or will not grow adjacent to the disk. The characteristic gram-positive diplococcus morphology of *Streptococcus pneumoniae* should always be confirmed (how?).

**Beta-hemolytic streptococci.** The beta-hemolytic streptococci are *all* pathogenic and must be carefully identified. One of the most important differential tests is the sensitivity of group A streptococci to the antibiotic Bacitracin. This sensitivity is measured by placing a disk with the antibiotic on blood agar that has been streaked with the beta-hemolytic isolate. Further differentiation of the other groups (B, C, and D) of beta-hemolytic streptococci requires the determination of several biochemical characteristics. Consult the Resource References for additional tests as needed.

**Enterococci.** Many, but not all, of the enterococci are nonhemolytic. These organisms are part of the normal flora of the intestinal tract, and like the alpha-hemolytic streptococci from the throat, can be important pathogens when introduced into other parts of the body. The enterococci can be distinguished from all of the other streptococci, regardless of hemolytic characteristics, by their growth in the presence of bile salts, growth in 6.5% salt, and the hydrolysis of esculin. Since the enterococci may exhibit alpha- and beta-hemolysis, the presence of group D enterococci must be considered, especially from urogenital-tract cultures and from heavily (4+) infected upper respiratory cultures. These three tests are usually considered sufficient to confirm the presence of one of the enterococci, regardless of the type of hemolysis observed.

**Neisseria.** Since up to half the colonies that appear from a throat culture can be members of the genus *Neisseria*, selective media must be used to isolate the two pathogenic species: *Neisseria gonorrhoeae* and *Neisseria meningitidis*. Both of these organisms will grow well on both Martin-Lewis and Thayer-Martin media, while all other neisseriae are usually inhibited. Growth on either of these media, when confirmed as being oxidase-positive, gram-negative diplococci, is considered presumptive for either of the species. Also, neither of them will grow on ordinary nutrient agar, while most of the nonpathogenic forms will. These growth characteristics are used as part of the diagnostic protocol for the neisseriae.

**Diphtheroids.** The diphtheroids, like the neisseriae, are among the most prominent isolates from healthy throats. As with the neisseriae, identification of the pathogen *Corynebacterium diphtheriae* requires a special isolation medium and additional biochemical tests. Recognition of the typical gram-positive-rod cellular morphology of the diphtheroids (Exercise 3) and the formation of metachromatic granules is critical for the correct identification of these organisms. You should consult the Resource References for any additional information you need.

## LABORATORY OBJECTIVES

In this exercise you will obtain a throat culture from one or more of your classmates and compare the growth obtained with known cultures provided to you. This exercise will require that you:

1. Take at least one throat culture.
2. Examine the growth that develops on the media used and attempt to recognize colonial characteristics typical of those media.
3. Acquire an understanding of the proper protocol for the identification of the organisms typically encountered in normal and abnormal throat cultures.
4. Acquire an understanding of the nature of the normal flora of the throat and how it is distinguished from the pathogenic forms that might colonize that area.
5. Develop an understanding of the routine diagnostic tests used to characterize the organisms used in this exercise.

## MATERIALS NEEDED FOR THIS LABORATORY

1. Sufficient number of plates to take the required number of throat cultures and to make streak plates of the organisms used as reference cultures.

    Blood agar (BA)
    Chocolate agar/Thayer-Martin biplates (CA/TM)
    Mannitol-salt agar (MSA)
2. Twenty-four-hour broth cultures of reference bacteria (see the next section).
3. Additional plates of media as needed to complete confirming tests.
4. Sterile swabs.
5. Reagents to complete screening tests:
    Hydrogen peroxide
    Coagulase plasma
    Bile-salt solution or disks (e.g., Taxo A, Taxo P, Optochin)
    Bacitracin disks
    Oxidase reagent or Taxo N disks
6. Additional media and reagents as required for supplemental diagnostic tests as directed by your instructor.

## LABORATORY PROCEDURE

1. Obtain and label the required number of plates according to the following:
   a. For each throat culture, one plate each of the following media:
       Blood agar
       Chocolate/Thayer-Martin
       Mannitol-salt agar
   b. One blood agar plate for each of the streptococci.
   c. One chocolate/Thayer-Martin plate for each of the neisseriae.
   d. One mannitol-salt agar plate for each of the staphylococci.
2. Make streak plates of all the known cultures on the appropriate agar. Because some of these bacteria are highly pathogenic, your laboratory instructor may demonstrate some of them for you. The cultures may include:

    *Branhamella catarrhalis*
    \**Corynebacterium diphtheriae*
    \**Neisseria gonorrhoeae*
    \**Neisseria meningitidis*
    *Staphylococcus aureus*
    *Staphylococcus epidermidis*
    *Streptococcus faecalis*
    \**Streptococcus pneumoniae*
    \**Streptococcus pyogenes*
    Alpha-hemolytic *Streptococcus*

*Note:* Your laboratory instructor may demonstrate the reactions of the pathogenic species (indicated with *) used in this exercise. In any case, be extremely cautious and be especially diligent in employing aseptic techniques.

3. Set up at least one throat culture from one of your classmates or from any individual who wishes to volunteer.
4. Incubate all cultures for 24 hours at 37°C.
5. Examine all plates and record the colony morphology of at least five types in addition to the reference cultures.
6. Complete screening and/or confirming tests on each of the known cultures and any suspect colonies observed on the throat cultures.
7. After you have completed your examination of the blood agar and chocolate agar plates, you may flood them with oxidase reagent. After a few minutes, oxidase-positive colonies will turn pink and then gradually turn dark purple. If the colonies are transferred while they are still pink, they may produce viable cells. After the purple color is evident, the cells will probably be dead.

## REFERENCES

### Text References

1. Atlas, Ronald M., *Microbiology: Fundamentals and Applications.* Chapters 13, 16, 17, and 18.
2. Brock, Thomas D., David W. Smith, and Michael T. Madigan, *Biology of Microorganisms,* 4th ed. Chapters 14, 15, 17, and 18.
3. Nester, Eugene W., et al., *Microbiology,* 3rd ed. Chapters 13, 22, and 23. Appendixes V and VII.
4. Stanier, R. Y., E. A. Adelberg, and J. Ingraham, *The Microbial World,* 4th ed. Chapters 23 and 30.
5. Stanier, R. Y., et al., *Introduction to the Microbial World.* Chapters 12 and 17.
6. Wistreich, George A., and Max D. Lechtman, *Microbiology,* 4th ed. Chapters 27, 29, and 30.

### Resource References

1. Finegold, Sidney M., and W. J. Martin, *Bailey and Scott's Diagnostic Microbiology,* 6th ed. Chapters 5, 6, 8, 14, and 16–32 (consult specific references).
2. Lennette, E. H., ed., *Manual of Clinical Microbiology,* 3rd ed. Chapters 2 and 6.

## LABORATORY PROTOCOL

### Exercise 11—Throat Cultures

Check each step when you complete it.

### First Day

_____ 1. Streak out each reference organism on the media shown. Streak for isolation.

_____ 2. Add the appropriate disks (Taxo P, Optochin, Taxo A, or Bacitracin) to the plates that have been inoculated with organisms that the disks are diagnostic for:

        Alpha-streptococcus:    Taxo P; Optochin
        Beta-streptococcus:     Taxo A; Bacitracin

_____ 3. Set up throat cultures as directed. Use a full range of selective and differential media.

_____ 4. Incubate all plates at 37°C for at least 24 hours, but not more than 48 hours.

### Second Day

_____ 5. Study and record colonial characteristics of all reference cultures (including any demonstrations). Note the following for each genus on each medium:

    Color of colony
    Diameter of colony
    Color of agar around colony
    Type of hemolysis, if any
    Growth/no growth on TM/ML or around any disks.

_____ 6. Study the throat culture plates and record the following:

    a. Presence of any beta-hemolytic colonies on blood agar. These should be picked and tested to determine if they are in group A.

    b. Approximate proportion of alpha-*Streptococcus, Neisseria,* and diphtheroids. Report as 1+, 2+, 3+, or 4+.

    c. When there is heavy growth of alpha-streptococci in the 3+ area, the colonies must be tested to determine if they are *Streptococcus pneumoniae*. This would be very important if the culture came from the very young or the elderly.

    d. Any growth on Thayer-Martin agar. Any colonies that do appear must be gram stained and tested with the oxidase reagent.

    e. Any growth on mannitol-salt agar must be tested with catalase and coagulase to determine if it is *Staphylococcus aureus*.

    f. Gram stain and observe all suspect and reference cultures.

_____ 7. Flood blood agar and chocolate agar plates with oxidase reagent; observe any color changes in the colonies.

_____ 8. Write a short report on the results of the throat culture, summarizing the findings from step 6.

_____ 9. Briefly describe the characteristics of each reference culture, summarizing the findings from step 5.

# NOTES

# DICHOTOMOUS KEY—BACTERIA COMMONLY ENCOUNTERED IN THROAT CULTURES

I. Gram negative
    A. Grows on Thayer/Martin agar, but not on nutrient agar
        1. Ferments glucose and maltose .................................... *N. meningitidis*
        2. Ferments glucose, but not maltose ................................. *N. gonorrhoeae*
    B. Grows on nutrient agar, but not on Thayer/Martin agar
        1. Ferments at least one sugar ........................................ *Neisseria sp.*
        2. Does not ferment any sugar ....................................... *B. catarrhalis*

II. Gram positive
    A. Catalase positive
        1. Aerobic ........................................................ *Micrococcus sp.*
                                                                                                               (Not *Staphylococcus*)
        2. Facultative
            a. Coagulase positive, ferments mannitol ....................... *S. aureus*
            b. Coagulase negative, does not ferment mannitol ............. *S. epidermidis*
    B. Catalase negative
        1. Grows in 6.5% salt and in medium containing bile .............. *Streptococcus*
                                                                                                                (Enterococcus)
        2. Alpha-hemolytic
            a. Soluble in bile solution ..................................... *S. pneumoniae*
            b. Not soluble in bile solution .................................. *Streptococcus*
                                                                                                               (Not *S. pneumoniae*)
        3. Beta-hemolytic
            a. Bacitracin sensitive ........................................ *Streptococcus*
                                                                                                  (Group A)
            b. Bacitracin resistant ....................................... *Streptococcus*
                                                                                     (Group B, C, or D)

Name: _____

## LABORATORY REPORT FORM

**Exercise 11—Throat Cultures**

1. Report the results for each of your throat cultures according to the following form (use 1+, 2+, 3+, and 4+):

| NOTES | RESULTS |
|---|---|
|  | Name of Patient: _____ |
|  | Date: _____ |
|  | *Blood Agar Results* |
|  | Number of alpha-streptococci: _____ |
|  | Number of beta-streptococci: _____ |
|  | Number of neisseria: _____ |
|  | Number of staphylococci: _____ |
|  | Bile solubility of streptococci: _____ |
|  | Coagulase test of staphylococci: _____ |
|  | *Chocolate/Thayer-Martin Results* |
|  | Number of colonies on T-M agar: _____ |
|  | Gram reaction of these: _____ |
|  | Number colonies on the chocolate agar: _____ |

## ANSWER THE FOLLOWING QUESTIONS

1. What two tests can be used to screen for bile solubility?

    a.

    b.

2. What is the significance of the bile solubility test?

3. How would you differentiate between *Staphylococcus aureus* and *Streptococcus pneumoniae*?

4. Why is it important to be able to rapidly differentiate between *Staphylococcus aureus* and *Staphylococcus epidermidis*?

5. Thayer-Martin agar inhibits almost all bacteria except:

   a.

   b.

6. What are the three types of hemolytic reactions demonstrated by the streptococci?

   a.

   b.

   c.

7. Name some pathogenic bacteria from each of the hemolytic groups.

# Exercise 12
# Urinary Tract Cultures

## SEMI-QUANTITATIVE (CALIBRATED-LOOP) ESTIMATION OF NUMBERS OF BACTERIA IN URINE

Urine from a healthy person, when carefully taken as a clean midstream or catheterized sample, will be found to contain remarkably few bacteria. One of the indicators of a pathological condition in the urinary tract is the presence of more than 100,000 bacteria/ml of urine. Most laboratory protocols for the analysis of a urine culture call for some kind of quantitative measure of the bacterial density and identification of the significant potential pathogens.

A clean midstream urine sample is one that is not contaminated with normal flora of the lower urogenital tract or the external genitalia. It must be carefully obtained. The patient is asked to cleanse the external genitalia and to void a portion of urine before collecting the sample. The initial voiding is necessary to rinse bacteria and cellular debris from the urethra.

The traditional method of determining the bacterial count is to perform a standard plate count (Exercise 24) on the urine sample. This is accomplished by making pour plates of a series of tenfold serial dilutions of the urine. Since the laboratory usually is only interested in determining if the count exceeds 100,000 bacteria/ml, pour plates of dilutions up to only 1:1,000 are usually made. Pour plates, however, are not suitable for rapid detection of pathogens; it is also necessary to make isolation streak plates of the sample on differential or selective media. The sample may either be streaked out directly or after mild centrifugation to concentrate the cells in a smaller volume of urine.

A newer and more commonly used alternative procedure to the pour plate is a quantitative technique that allows for the simultaneous estimation of numbers and the direct isolation of pathogens on diagnostic and selective media. This alternate technique is known as the *calibrated-loop procedure*. A calibrated loop is one that has been carefully calibrated to hold a drop of urine of a specific volume, usually 0.001 ml. The loop is dipped into the urine sample and a single streak is made down the center of the plate being used for isolation. This streak is then cross-streaked for isolation of any bacteria that may have been deposited. Any medium may be used, allowing simultaneous isolation of pathogens and estimation of numbers.

Studies have established that when the loops are properly calibrated, the numbers of bacteria in the urine are proportional to the extent of growth down the center streak. If the growth extends more than three fourths of the distance, the urine is considered to have more than 100,000 bacteria/ml and is considered to be evidence of a pathogenic condition. When this occurs, the dominant organism, as well as any suspected pathogens, must be isolated and identified. A more precise estimate can be obtained by counting the colonies that appear on the streak and multiplying that number by 1,000 (0.001 ml was plated) to give the number of bacteria/ml.

## ORGANISMS FREQUENTLY ENCOUNTERED

The most commonly encountered bacteria from urine cultures include the gram-negative enteric bacteria, the streptococci, and the staphylococci. *Neisseria, Lactobacillus,* and yeasts are also often isolated. The importance of obtaining a good, clean, midstream sample cannot be overemphasized. One of the most common problems encountered with urine cultures is contamination resulting from improper preparation before sample collection and the failure to correctly obtain a clean, midstream sample. This is a greater problem with female patients than with males, for obvious reasons. Failure to carefully cleanse the area, and to take the proper midstream sample could result in a sample that is contaminated with normal flora of the skin and associated genital organs (*S. epidermidis, Lactobacillus,* and diphtheroids) and could result in a course of therapy that is not necessary.

1. The *lactobacilli* are slender, gram-positive rods that produce lactic acid as a product of fermentation. They are catalase negative and are part of the normal flora of the female urogenital tract, particularly the vagina.
2. *Yeasts* can be identified easily by their typical large size (they are eucaryotic), their characteristic formation of daughter cells by budding, and occasionally by their "yeasty" odor. Since several of the yeasts can grow on Thayer-Martin agar and do produce a weak oxidase reaction, it is imperative to do a gram stain on *all* Thayer-Martin colonies. *Candida albicans* is a pathogenic yeast that can be identified relatively easily by the germ-tube test, which is described in the Resource References.
3. The *gram-negative rods* that infect the lower urogenital tract are usually introduced from the intestinal tract. Since we will cover this group in detail in the next exercise, you should save any gram-negative rods that are isolated. They may be initially screened by noting if they ferment lactose and are oxidase positive.

## MEDIA USED AND THEIR REACTIONS

**Blood agar.** Many of the streptococci isolated will be enterococcus and will appear to be nonhemolytic (gamma-hemolysis). If the sample contains large numbers of other bacteria, the detection of the streptococci may be difficult. Refer to Exercise 11 for a complete description of the reactions on blood, chocolate, and mannitol-salt agars.

**Chocolate agars.** The chocolate agars, including Thayer-Martin agar, are used to detect pathogenic neisseriae. Although most bacteria can grow well on chocolate agar, the selectivity of Thayer-Martin agar will ensure isolation of *N. gonorrhoeae* should it be present.

**Mannitol-salt agar.** This agar is occasionally used for the detection of *S. aureus,* although when *S. aureus* is present it can usually be detected on the blood agar streaks.

**MacConkey's agar.** MacConkey's agar is a medium that is used for the selection and differentiation of the enteric gram-negative rods associated with the intestinal tract. It contains crystal violet to suppress the growth of gram-positive bacteria. Other components of the medium, including the sugar lactose and a pH indicator (neutral red; red at 6.8, yellow at 8.0), allow differentiation between lactose-fermenting and lactose-nonfermenting gram-negative rods. Lactose-positive bacteria will produce red colonies with a distinctly darker red zone around the colony (due to the color change of the pH indicator), while lactose-nonfermenting bacteria will produce colonies that are light pink or colorless after 24-hour incubation. Whenever the gram-negative rods comprise a significant proportion of the isolated colonies, identification is required, especially if the total count exceeds 100,000/ml.

## LABORATORY OBJECTIVES

This laboratory exercise will introduce you to new procedures that are commonly used in the bacteriological examination of urine, including quantitative techniques for the estimation of the numbers of bacteria in urine. To complete this exercise you will:

1. Understand and explain what a serial dilution is and what it is used for.
2. Explain and differentiate between pour plates and streak plates and discuss their advantages and disadvantages relative to each other.

3. Know what a calibrated loop is and what its applications are.
4. Obtain an understanding of the proper protocol for bacteriological analysis of urine samples.
5. Discuss the importance of a properly obtained urine sample.
6. Achieve some appreciation for the normal and pathogenic flora of the lower urogenital tract.

## EXPERIMENTAL DESIGN

This exercise includes both a pour plate and calibrated-loop streak plate. Unless your instructor indicates otherwise, you should complete both procedures. Also, you will be expected to complete any additional screening test to identify specific genera or groups of bacteria. For example, you may be required to perform the catalase test to confirm the isolation of *Streptococcus* and the bile-esculin test to confirm that the *Streptococcus* is an enterococcus.

## MATERIALS NEEDED FOR THIS LABORATORY

*Note:* You may not be required to complete all of the procedures listed in the next section of this exercise, and the materials you will need to complete the exercise may vary accordingly. For example, your laboratory instructor may have some groups do pour-plate counts and others do the calibrated-loop count.

1. Standard plate count of a urine sample requires four plates for each sample (one each for the undiluted sample and for each of these dilutions: 1:10, 1:100, and 1:1000).
2. Dilution blanks, typically three for each sample.
3. Four plate-count agar deeps, containing 20 ml of melted agar, held at about 50–52°C.
4. Sterile pipettes, 1.0 ml.
5. Calibrated loops.
6. Differential and selective medium plates for isolation streak plates of urinary tract samples and for streak plates of known reference cultures:
   Blood agar/MacConkey's agar (biplates recommended)
   Chocolate agar/Thayer-Martin agar (biplates recommended)
   Mannitol-salt agar plates
7. Reagents and media to complete screening tests, including:
   Catalase
   Coagulase
   Oxidase
   Gram-stain reagents
   Bile/esculin agar
8. Urine samples. Urine samples may be provided for you, or you might be asked to obtain your own (collection bottles and instruction will be provided). Ideally, each group should analyze one normal sample and one pathogenic sample.

## LABORATORY PROCEDURE

1. Obtain and label the following differential media for each urine sample (use biplates, if available):
   Blood agar
   MacConkey's agar
   Chocolate agar/Thayer-Martin agar
   Mannitol-salt agar
2. For a standard plate count, obtain each of the following:
   Plate-count agar deeps, four, melted, held at 50–52°C.
   Sterile petri plates, four, labeled according to dilution.
   Sterile dilution blanks, three, containing 9.0 ml of 0.9% saline, labeled according to dilution.
3. If you are counting by the calibrated-loop procedure, you can use the differential and selective media obtained in step 1.
   a. Follow the flow diagram in Fig. E12-1: Carefully clean the loop, following any special instructions for cleaning and flaming that might be provided by the instructor.
   b. Dip the loop into the urine sample so that the loop is just below the surface of the sample. No part of the stem should be in the urine (urine adhering to the stem will cause the counts to come out higher than they should).
   c. When you remove the loop from the urine, the loop will be filled with a

droplet of sample. If the loop has been properly calibrated and correctly handled, it will contain about 0.001 ml.
d. Make a single streak down the center of the plate by touching the filled loop to the top center and quickly sliding the loop down to the bottom center of the plate. Cross-streak over the first streak to isolate colonies. Use your regular loop or needle for this isolation streak.
e. Repeat this procedure for each type of diagnostic medium used.

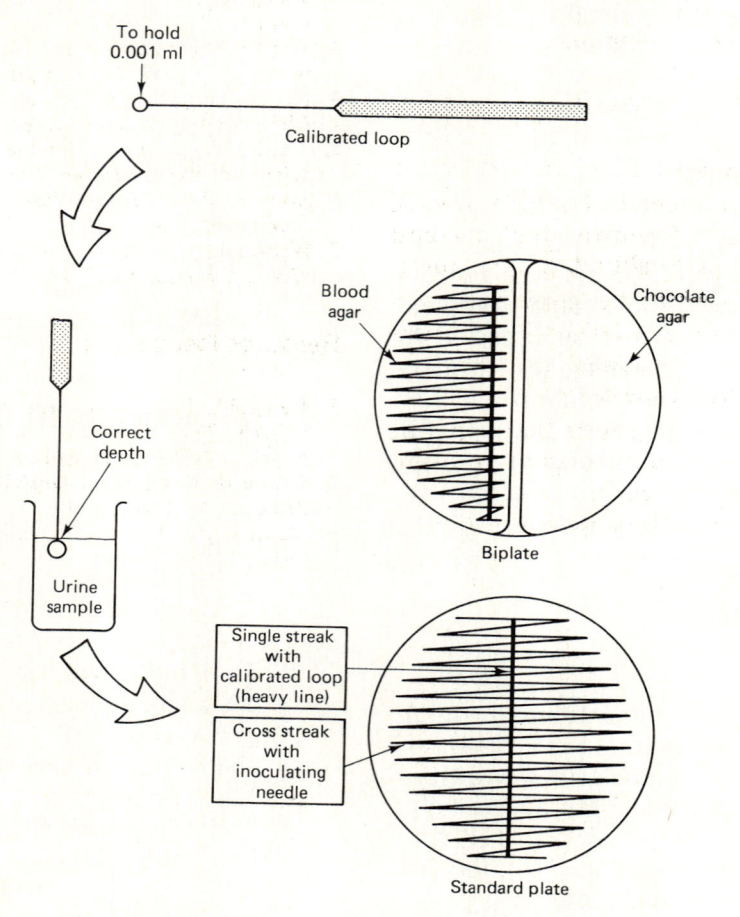

**Figure E12-1** Calibrated loop-streak plates.

4. Complete the standard plate count of the urine sample. Study the flow diagrams in Exercises 1 and 24:
   a. With sterile 1.0-ml pipettes, transfer 1 ml of urine into the first dilution blank *and* transfer 1 ml into the first tube of melted agar.
   b. Discard the pipette into the pipette jar provided for that purpose. Do not use any pipette more than once.
   c. Mix the agar by gently mixing or inverting the closed tube several times. *Do not shake vigorously.* Pour the seeded medium into the first plate and allow it to solidify.
   d. Mix the seeded dilution blank by vigorous shaking or swirling. Transfer 1 ml from the first blank to the second *and* 1 ml to the second tube of melted agar.
   e. As before, mix the medium gently, pour into a petri dish, and allow to solidify.
   f. Similarly, mix the second dilution blank and make the transfers to the next blanks and melted agar tubes.
   g. Continue until all four plates have been poured. You should have one plate each for undiluted urine (first plate), 1:10 (second plate), 1:100

(third plate), and 1:1000 (fourth plate).
5. Make isolation streak plates of any reference cultures provided. These may include:
   *Staphylococcus aureus*
   *Staphylococcus epidermidis*
   *Streptococcus faecalis*
   *Escherichia coli*
   *Pseudomonas aeruginosa*
   *Lactobacillus sp.*
   *Saccharomyces cerevisiae* or other yeasts
6. Incubate all plates at 37°C for 24 hours.
7. Determine the number of bacteria in each milliliter of urine. Compare the standard plate count with the calibrated-loop count.
8. Observe and describe the growth characteristics of the reference cultures and determine if any similar colonies developed on streak plates of the urine samples.
9. Complete any screening tests that might be needed (see the section entitled "Organisms Frequently Encountered").
10. Complete the Laboratory Report Form for this exercise.

## REFERENCES

### Text References

1. Atlas, Ronald M., *Microbiology: Fundamentals and Applications.* Chapters 13, 16, 17, and 18.
2. Brock, Thomas D., David W. Smith, and Michael T. Madigan, *Biology of Microorganisms,* 4th ed. Chapters 14, 15, 17, and 18.
3. Jensen, Marcus, *Introduction to Medical Microbiology.* Consult specific topical references.
4. Nester, Eugene W., et al., *Microbiology,* 3rd ed. Chapters 13, 22, and 23. Appendixes V and VII.
5. Stanier, R. Y., E. A. Adelberg, and J. Ingraham, *The Microbial World,* 4th ed. Chapter 9.
6. Stanier, R. Y., et al., *Introduction to the Microbial World.* Chapter 5.
7. Wistreich, George A., and Max D. Lechtman, *Microbiology,* 4th ed. Chapters 27 and 33.

### Resource References

1. Finegold, Sidney M., and W. J. Martin, *Bailey and Scott's Diagnostic Microbiology,* 6th ed. Chapters 5, 6, 10, and 16–32 (consult specific references).
2. Lennette, E. H., ed., *Manual of Clinical Microbiology,* 3rd ed. Chapters 2 and 6.
3. McGonagle, L. A., *Procedure for Diagnostic Microbiology.* Page 134.

## LABORATORY PROTOCOL

### Exercise 12 — Urinary Tract Cultures

Check each step when you complete it.

**First Day**

_____ 1. Set up a standard plate count on the urine sample(s) provided,

and/or

_____ 2. Set up a calibrated-loop count of the urine sample(s).

_____ 3. Make isolation streak plates of all reference cultures on the appropriate media.

_____ 4. Incubate all plates for 24 hours at 37°C.

**Second Day**

_____ 5. Record the growth characteristics of the reference bacteria on the media used, especially any new media (e.g., MacConkey's).

_____ 6. Complete the written reports as directed in the exercise.

# NOTES

# DICHOTOMOUS KEY—BACTERIA COMMONLY ENCOUNTERED IN URINARY TRACT CULTURES

I. Gram-positive bacteria
    A. Catalase positive
        1. Coagulase and gelatinase positive ............................. *S. aureus*
        2. Coagulase and gelatinase negative ......................... *S. epidermidis*
    B. Catalase negative
        1. Cocci ........................................................ *S. faecalis*
        2. Rods ..................................................... *Lactobacillus sp.*
II. Gram-negative bacteria
    A. Oxidase positive ............................................... *P. aeruginosa*
    B. Oxidase negative ..................................................... *E. coli*
III. Typical yeast morphology ............................................. *S. cerevisiae*

*Note:* This key is limited to those tests necessary for the identification of the bacteria used in this exercise (see page 127). Additional organisms would, of course, require additional tests. Also, a good microbiologist would insist on some confirmatory tests for each isolate.

Name: _____

## LABORATORY REPORT FORM

**Exercise 12 — Urinary Tract Cultures**

1. Complete the table shown below. (If you are not required to complete three analyses, obtain results from other students in the laboratory.)

### URINE COLONY COUNTS

| Sample Source: | | | |
|---|---|---|---|
| Pour-plate counts: | | | |
| Calibrated-loop counts: | | | |
| Blood agar | | | |
| Chocolate agar | | | |
| Thayer/Martin agar | | | |
| MacConkey's agar | | | |
| Mannitol-salt agar | | | |

2. Write a short description of the types of colonies that developed from each sample.

3. Describe the colony characteristics of each reference culture.

4. Describe the gram-stain characteristics of yeast and *Lactobacillus*.

## ANSWER THE FOLLOWING QUESTIONS

1. Explain how a clean, midstream or clean-catch sample is obtained.

2. List four commonly encountered pathogens associated with lower urinary-genital-tract infections.

3. Explain the color reactions observed on MacConkey's agar.

4. Describe the characteristics of the enterococcus on MacConkey's agar.

5. Discuss any apparent differences between the pour-plate counts and the calibrated-loop counts.

# Exercise 13
# Gastrointestinal Tract Cultures

## GRAM-NEGATIVE ENTERIC BACTERIA

Bacterial diseases of the gastrointestinal tract can take many forms—from moderately discomforting "stomach flu" to severe diarrhea that could result in life-threatening loss of fluids and electrolytes. Almost every genus of the gram-negative enteric bacteria have species that can be pathogenic. Two of the more notable ones are *Salmonella* and *Shigella,* but even *Escherichia* and *Enterobacter* may cause intestinal disease under the proper circumstances. Of course, all of these bacteria, whether usually considered pathogenic or not, are to be considered opportunistic pathogens that will result in severe bacterial infections when introduced to parts of the body that they are not usually associated with.

Some nonenteric gram-negative rods, such as certain species of *Pseudomonas* and *Vibrio,* are frequently encountered with the enteric bacteria. One of them, *Pseudomonas aeruginosa,* is usually thought of as nonpathogenic, but can be a very serious pathogen in immunologically compromised hosts and in burn patients. *P. aeruginosa* is metabolically very versatile and resistant to many of the commonly used antibiotics. It produces a characteristic green pigment in many kinds of growth media. *Vibrio* is a genus of gram-negative, curved rods. Most of the species in the genus are saprophytes, especially in marine environments. Two notable exceptions include *V. cholerae,* which causes the disease of the same name, and *V. parahaemolyticus,* a species that is associated with food poisoning involving seafoods. Both *Pseudomonas* and *Vibrio* are oxidase positive and lactose negative.

The intestinal tract also provides an ideal growth environment for many other kinds of bacteria, including several kinds of enterococci and many anaerobes. Such a large and diverse assembly of bacteria presents some unique problems when it becomes necessary to detect pathogenic species that might be present in very small numbers. Several procedures using highly selective media and isolation or enrichment media have been developed to help resolve this problem.

## STRATEGIES FOR STOOL CULTURES

**Enrichment cultures.** When some members of a bacterial population are present in only a very small proportion of that population, it is necessary to use enrichment media for their isolation. The medium chosen is one that contains selective inhibitors that will inhibit the growth of all bacteria except the species that is to be isolated. The population is said to be *enriched*—that is, the population is altered by favoring the growth of one kind of bacterium at the expense of the others. The objective of the procedure is to increase the likelihood of isolating the pathogenic species on a streak plate—something that would not be possible without first enriching the population for the pathogen.

**Selective media.** Most of the media used for isolation of the enteric bacteria contain inhibitors to suppress the growth of nonenteric organisms. The two most commonly used inhibitors are the bacteriostatic dyes and bile salts. The dyes inhibit the growth of gram-positive bacteria, and the bile inhibits those bacteria that are not normally found in the intestinal tract (why?).

**Differential media.** Many of the differential media used to identify the gram-negative rods do so on the basis of fermentation reactions. When acid is produced, a pH indicator in the medium changes color; a good rule of thumb to follow is that if the colony is a different color than the medium, then it ferments the sugar. Most of these media use lactose or sucrose as the indicator sugar because virtually none of the pathogenic forms (with one or two notable exceptions) ferment one or the other of these sugars. Therefore, using the rule of thumb noted above, any colony that remains the same color as the medium should be considered suspect and identified. Several media also use additional selective or differential devices; specific information about them can be obtained from the Reference Resources. Remember that the media commonly used today are both selective and differential.

**Multitest media.** In addition to the plated media, there are several tubed media that have differential properties and can be used to perform more than one test on an isolate. One of the most common is *triple sugar iron agar* (*TSI*). It contains three sugars and a pH indicator. The sugars are lactose, sucrose, and glucose, but the lactose and sucrose concentration is ten times (10×) greater than the glucose concentration. When this medium is correctly inoculated, bacteria that ferment glucose only will cause the medium to turn yellow in the bottom of the tube and a red color at the top (why?). Bacteria that ferment either lactose or sucrose will cause a yellow color to develop throughout the tube. If gas is also produced, bubbles will appear in the agar, causing it to "break" into pieces. This medium also contains iron salts that will turn black when hydrogen sulfide is produced. TSI can, therefore, be used to determine fermentation reaction of three sugars (but not all three at once) and can test for the production of hydrogen sulfide. Several other multitest media will be used in this exercise; their reactions are explained in Appendix 2.

**Summary.** A typical strategy for the bacterial analysis of a stool culture (or any culture that may contain gram-negative rods) is as follows:

1. Enrichment: Use of a medium that will enhance the growth of the bacterium you want to isolate so that it becomes a prominent member of the population and can be more easily isolated on a streak plate.
2. Selection and differentiation: Use of appropriate media that inhibit the growth of nonenteric bacteria and which differentiate between those that ferment lactose or sucrose and those that do not.
3. Biochemical identification: Identification on the basis of biochemical reactions, including those determined on certain multi-test media.

## MEDIA USED AND THEIR REACTIONS

It is especially important that you study the growth patterns and colony morphologies of the reference cultures on the media listed below. Subtle differences in color, colony margin, consistency, and so on can be very important diagnostic leads.

**1. Tetrathionate broth.** This is a highly selective medium that is used to enhance the growth of the genera *Salmonella* and *Shigella*. The liquid medium is inoculated with a swab saturated with the stool sample and incubated for 12 to 24 hours. Selenite F is another example of this type of selective medium.

**2. Alkaline peptone medium.** Alkaline peptone medium (APM) is used to enrich samples for *Vibrio*. It contains a high salt concentration (3.0%) and a high pH (about 8.0).

**3. MacConkey's agar.** This medium was discussed in the previous exercise. The combination of bile salts and bacteriostatic dyes makes the medium selective for the enteric bacteria. Lactose and pH indicators allow for differentiation between lactose-fermenting and nonfermenting rods.

4. **Hektoen enteric agar.** This medium is highly selective and differential. Organisms that ferment lactose will produce yellow to pink colonies, while the nonfermenters will be smaller and green in color. This medium is very reliable and is one of the most commonly used today.

5. **SS agar.** Salmonella-Shigella agar is selective for these two genera. However, like most such media, the selectivity is not absolute (*Pseudomonas* grows fairly well on it) and gram stains and confirming biochemical tests are always required. *Salmonella* and *Shigella* colonies are light pink, with *Salmonella* occasionally producing a colony with dark centers. The occasional lactose fermenter that does grow on SS agar will produce a distinctly more red colony.

6. **Triple sugar iron agar.** TSI was discussed previously. A yellow color indicates fermentation of one of the sugars. When only glucose is fermented, the deepest part of the medium (the *butt*) will be yellow, while the part of the medium that is exposed to air (the *slant*) will be red. The red color appears when glucose is depleted, forcing the bacteria to use the protein components of the medium for growth. When either, or both, of the other sugars is used, the entire tube will be yellow because the concentration of the sugar is large enough so it does not become depleted. As noted above, the iron will react with any hydrogen sulfide produced and cause the formation of a black precipitate in the deep part of the butt. Another medium, Kligler's iron agar, is very similar to TSI except that it has only two sugars (glucose and lactose).

7. **SIM agar.** SIM agar is a medium that can be used to determine motility (M), sulfide (S) production, and indole (I) production. The medium is soft enough so motile bacteria can swim through it, producing a very hazy and cloudy central stab. Nonmotile bacteria produce a sharp and clear stab (they can't swim). The sulfide is detected by formation of a black precipitate, and the indole is detected by the addition of Kovac's reagent (see previous exercises).

8. **TCBS agar.** TCBS stands for *thiosulfate citrate bile sucrose agar*. It is highly selective for the genus *Vibrio* because of the high pH (about 8.0) and high salt content. It differentiates between sucrose-fermenting and nonfermenting bacteria. The medium can be used for the isolation of *Vibrio* cholera (yellow colonies) and *V. parahaemolyticus* (green colonies). It is often included in stool-culture protocols during the summer months, when *V. parahaemolyticus* is most likely to be encountered.

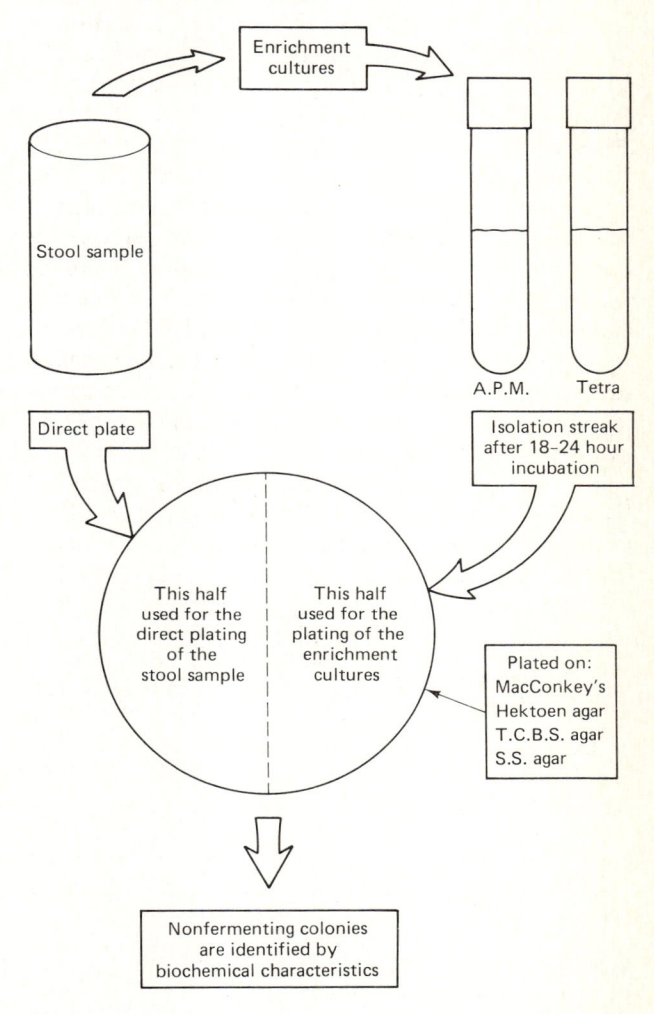

**Figure E13-1** Stool-culture protocol with enrichment for *Vibrio* and *Salmonella*/Shigella.

## LABORATORY OBJECTIVES

Stool cultures are unique because, more so than with any other type of culture, the selection and enrichment for pathogens is an important part of the procedure. In this exercise you will need to:

1. Understand the enrichment process and why it is needed.

2. Understand the principles involved in the use of selective and differential media.
3. Correctly identify lactose- and sucrose-fermenting bacteria on the appropriate media.
4. Understand the procedures and strategies employed in the analysis of stool cultures.

## RECOMMENDED PROTOCOL FOR STOOL CULTURES

**Day 1.** Inoculate at least one enrichment medium and at least three selective and differential media by streak plate. Some workers recommend that half of the plates be used for the initial streak, the other half used for streaking the enrichment cultures. The recommended media include:

a. Enrichment media:
  Selenite F or tetrathionate broth
  Alkaline peptone medium
    (for *Vibrio*)
b. Isolation media:
  MacConkey's agar
  TCBS agar (for *Vibrio*)
  Hektoen enteric agar
  SS agar

**Day 2.** Examine the plates previously inoculated and make streak plates of the enrichment culture.

**Day 3.** Examine all plates for the presence of lactose nonfermenting bacteria. Also examine the TCBS plates for typical *Vibrio* colonies (yellow or green). Examine all reference cultures. Initiate any biochemical identifications that are appropriate, including TSI, SIM, and/or KIA media.

## THE IMViC TEST

The IMViC test is commonly used in the bacterial analysis of water to differentiate rapidly between *Escherichia coli* and *Enterobacter aerogenes*. It is not as commonly used in clinical laboratories because other screening tests have proven more useful. Nevertheless, virtually all screening tests for the gram-negative rods use these reactions (in addition to others) in their protocol.

|  | I | M | V(i) | C |
|---|---|---|---|---|
| Test | Indol | Methyl Red | Voges-Proskauer | Citrate |
| E. coli | + | + | − | − |
| E. aerogenes | − | − | + | + |

## MATERIALS NEEDED FOR THIS LABORATORY

1. For each stool culture:
    Tetrathionate or Selenite F broth
    Alkaline peptone medium
    MacConkey's agar
    TCBS agar
    Hektoen enteric agar
    SS agar
2. For each reference culture and for each isolate:
    TSI agar tubes
    SIM agar tubes
3. Recommended reference cultures:
    *Escherichia coli*
    *Enterobacter aerogenes/cloacae*
    *Proteus vulgaris*
    *Pseudomonas aeruginosa*
    *Proteus mirabilis*
    *\*Vibrio parahaemolyticus*
    *\*Salmonella typhimurium*
    *\*Shigella sp.*

    *Note:* Some of the bacteria (indicated with \*) you will study in this exercise are pathogenic. Your laboratory instructor may use demonstration cultures of these pathogenic forms. In any case, use extreme caution and pay particular attention to your aseptic techniques.
4. Sufficient biochemical media to complete screening and/or identification as directed by the instructor. Gram-stains reagents.

## LABORATORY PROCEDURE

1. Inoculate the following media with each stool sample provided:
   a. Selenite-F or tetrathionate broth
   b. Alkaline peptone medium
   c. MacConkey's agar
   d. TCBS agar
   e. Hektoen enteric agar
   f. SS agar

   The broth is heavily inoculated with a swab saturated with the stool sample. Half of each plate should be used for the primary isolation and the other half used for streaking the enrichment cultures.

2. Inoculate each of the media listed above with the reference cultures.
3. Incubate all cultures for 24 hours at 37°C.
4. Streak out the enrichment culture on the other half of the plates used in step 1.
5. After incubation, examine all streak plates, study and compare all colony types, and determine if any lactose nonfermenting bacterial colonies developed on the stool-culture streak plates.
6. Inoculate any suspect colonies and all the reference cultures into TSI or KIA media and into SIM agar. Also inoculate any biochemical tests needed to identify the isolates, as directed by your instructor.
7. Write a report summarizing your analysis of the stool cultures. It should include:
   a. A description of the reference cultures on each of the media used.
   b. A discussion on the use of the enrichment procedure, including the relative number of lactose nonfermenting bacteria before and after enrichment.
   c. A summary of the stool culture analysis: presence or absence of potential pathogens and protocol used for identification or screening.

## REFERENCES

### Text References

1. Atlas, Ronald M., *Microbiology: Fundamentals and Applications.* Chapters 13, 14, 17, and 18.
2. Brock, Thomas D., David W. Smith, and Michael T. Madigan, *Biology of Microorganisms,* 4th ed. Chapters 14, 15, 17, and 18.
3. Jensen, Marcus, *Introduction to Medical Microbiology.* Chapters 25 and 26.
4. Nester, Eugene W., et al., *Microbiology,* 3rd ed. Chapters 13, 22, and 23. Appendixes V and VII.
5. Stanier, R. Y., E. A. Adelberg, and J. Ingraham, *The Microbial World,* 4th ed. Chapter 29.
6. Stanier, R. Y., et al., *Introduction to the Microbial World.* Chapters 16 and 17.
7. Wistreich, George A., and Max D. Lechtman, *Microbiology,* 4th ed. Chapters 27 and 32.

### Resource References

1. Finegold, Sidney M., and W. J. Martin, *Bailey and Scott's Diagnostic Microbiology,* 6th ed. Chapters 5, 6, 9, and 16-32 (consult specific references).
2. Lennette, E. H., ed., *Manual of Clinical Microbiology,* 3rd ed. Chapters 16, 18, and 22.
3. McGonagle, L. A., *Procedures for Diagnostic Microbiology.* Pages 100-107, 139.

# LABORATORY PROTOCOL

**Exercise 13 — Gastrointestinal Tract Cultures**

Check each step when you complete it.

**First Day**

_____ 1. Inoculate the stool sample into an enrichment medium by placing a swab saturated with the sample directly into either tetrathionate broth or Selenite F broth.

_____ 2. Inoculate the stool sample directly onto differential and selective media. Make isolate streak plates on MacConkey's, Hektoen, TCBS, and SS agars. Use one half of the plates.

_____ 3. Make isolation streak-plate cultures of the reference organisms on the differential and selective media used in step 2. Use one half of a plate for each organism.

_____ 4. Incubate all cultures for 24 hours.

**Second Day**

_____ 5. Using the reference cultures as examples, examine the plates of the stool culture for colonies that do not ferment the indicator carbohydrate. If there are any, pick a colony and restreak to confirm that it is a pure culture.

_____ 6. Make streak-plate cultures from the enrichment cultures in tetrathionate, Selenite F broths, or APM. Use the other half of the plates from step 2.

_____ 7. Incubate all cultures for 24 hours.

**Third Day**

_____ 8. Examine the new streaks (made from the enrichment cultures) for suspect colonies. If any are found, pick them and streak onto another plate. Compare the two halves of the plates for any changes in the distribution of organisms.

_____ 9. Determine the oxidase reaction of all isolates and of the reference cultures.

_____ 10. Inoculate all cultures onto triple sugar iron agar (TSI) or Kligler's iron agar (KIA) and SIM agar.

_____ 11. Inoculate additional media, as needed, to complete any required identifications.

_____ 12. Incubate all cultures for 24 hours.

_____ 13. Record results and observations in the Laboratory Report Forms.

# NOTES

# DICHOTOMOUS KEY—BACTERIA COMMONLY ENCOUNTERED IN GASTROINTESTINAL TRACT CULTURES

I. Oxidase positive
    A. Ferments sucrose, curved rod .................................. *V. parahemolyticus*
    B. Does not ferment any sugar, produces green pigment in some media ....... *P. aeruginosa*
II. Oxidase negative
    A. Ferments lactose
        1. Produces indole ................................................ *E. coli*
        2. Does not produce indole ................................. *E. aerogenes*
    B. Does not ferment lactose
        1. Hydrolyzes urea
            a. Produces indole ................................... *P. vulgaris*
            b. Does not produce indole ............................. *P. mirabilis*
        2. Does not hydrolyze urea
            a. Motile ................................... *S. typhimurium*
            b. Nonmotile ................................. *Shigella sp.*

*Note:* This key is limited to those tests necessary for the identification of the bacteria used in this exercise (see page 138). Additional organisms would, of course, require additional tests. Also, a good microbiologist would insist on some confirmatory tests for each isolate.

Name: _____

# LABORATORY REPORT FORM

**Exercise 13 — Gastrointestinal Tract Cultures**

1. Complete the following chart. You will need to get some of the information from your laboratory partner and from other groups in the class.

### GROWTH CHARACTERISTICS OF SELECTED GRAM-NEGATIVE RODS

| Test Media | Colony Color |  |  |  | TSI and/or SIM |  |  |  |  |
|---|---|---|---|---|---|---|---|---|---|
|  | Mac | TCBS | Hek | SS | Butt | Slant | Sulfide | Indole | Motile |
| P. aeruginosa | ___ | ___ | ___ | ___ | ___ | ___ | ___ | ___ | ___ |
| P. vulgaris | ___ | ___ | ___ | ___ | ___ | ___ | ___ | ___ | ___ |
| P. mirabilis | ___ | ___ | ___ | ___ | ___ | ___ | ___ | ___ | ___ |
| Shigella sp. | ___ | ___ | ___ | ___ | ___ | ___ | ___ | ___ | ___ |
| E. coli | ___ | ___ | ___ | ___ | ___ | ___ | ___ | ___ | ___ |
| S. typhimurium | ___ | ___ | ___ | ___ | ___ | ___ | ___ | ___ | ___ |
| E. aerogenes | ___ | ___ | ___ | ___ | ___ | ___ | ___ | ___ | ___ |
| V. parahaemolyticus | ___ | ___ | ___ | ___ | ___ | ___ | ___ | ___ | ___ |

You should compare these results with the biochemical characteristics determined in Exercise 4. Be prepared to explain the color reactions on the petri plates and the colors observed in the butts and slants of the triple sugar iron agar.

2. Record the appearance of the streak plates before and after enrichment. Did the procedure increase the number of nonfermenting colonies?

## ANSWER THE FOLLOWING QUESTIONS

1. Explain why lactose and sucrose are the two most commonly used indicator sugars in the enteric isolation media.

2. Explain the enrichment procedure. What two enrichment media were used in this exercise?

3. Explain the term *reversion* with respect to the reactions on TSI agar.

4. What is bacterial endotoxin?

5. What is meant by *enteropathic E. coli*?

6. List one disease or pathogenic condition caused by each of the bacteria studied in this exercise.

# Exercise 14
# Lactobacillus Activity in Saliva

Tooth decay has been linked to bacteria found in the oral cavity. Bacteria that are part of the normal flora can form plaque and become lodged in lesions on the teeth. Through a combination of acid production by these bacteria and decomposition of the tooth structure, dental caries form and increase in size.

## THEORETICAL BASIS FOR THE FORMATION OF DENTAL CARIES

The formation of lactic acid by bacteria belonging to the genera *Streptococcus* and *Lactobacillus* has been directly linked to the formation of dental caries. These bacteria, growing in saliva and on the surface of the teeth, ferment available carbohydrates to produce the lactic acid. Most of these bacteria are able to ferment sucrose (table sugar) and complete its conversion to lactic acid in the plaque deposits on the teeth or in the saliva of the mouth. This fermentation reaction is completed before the sucrose-containing food is completely chewed and swallowed. (Those who have inflamed gums or teeth with dental caries will feel pain as the acids produced by these bacteria irritate the lesions.) The lowered pH increases the solubility of the enamel, resulting in a dental caries. The formation of plaque and the lodging of food and bacteria in tooth lesions create anaerobic conditions that enhance acid production through fermentation. The resulting process is, therefore, self-sustaining once it begins.

The mechanism of bacterial adhesion to the tooth surface and the formation of glycocalyx is explained in your text and will not be discussed here. It is, however, an important mechanism for the formation of plaque.

## EXPERIMENTAL DESIGN

Two procedures have been developed to estimate the amount of acid production by the bacteria in the oral cavity. Both procedures require that saliva be added to selective medium and acid production measured by color changes in the medium.

One approach is to actually count the number of bacteria present in saliva which produce acid. The other approach is to assume that acid production will be proportional to the number of bacteria and then to merely determine the rate at which the acid is produced; that is, the more acid, the more bacteria.

In the Rogosa test, direct pour-plate counts of the bacteria in saliva are made using Rogosa SL agar. This medium is highly selective for *Streptococcus* and *Lactobacillus*. The number of colonies that develop is assumed to be representative of the bacterial activity in the oral cavity.

In the Snyder test, saliva is added directly to a tube of Snyder test agar and incubated for up to 72 hours. The medium will turn from green to yellow as bacteria produce lactic acid. Since the production of acid is proportional to the number of bacteria, the sooner the medium turns yellow, the more bacteria present in the saliva sample.

## COLOR CHANGES AND NUMBER OF BACTERIA PER ML OF SALIVA

| Time of Color Change (Snyder Test) | Susceptibility to Dental Caries | Approximate Number of Bacteria per ml (Rogosa Test) |
|---|---|---|
| <72 hours | low | <100/ml |
| <48 hours | high | <10,000/ml |
| <24 hours | very high | >10,000/ml |

Both of these tests, in theory, measure the same thing—the amount of lactic acid produced by bacteria in saliva. In the Snyder test, an activity (production of acid) of the bacteria is measured, while in the Rogosa test, the actual number of acid-producing bacteria in saliva is determined. Both procedures assume that there is some direct relationship between the activity or numbers measured under laboratory conditions and the actual production of acid on the surfaces of teeth.

The value of these tests lies in the relationship that exists between the production of acid by bacteria and the formation of dental caries. The accuracy and reliability of the tests depend upon correct sampling, accuracy in reading the tests, and an understanding of the limitations of each procedure. Some important variables include such things as the timing of the sampling period (before or after brushing and flossing), early or late in the day (food accumulation), and the nature of the diet (high in sucrose). Also, variation among individuals, especially in their susceptibility to dental caries, often influences test results.

## LABORATORY OBJECTIVES

This exercise uses two approaches to determine susceptibility of individuals to dental caries. It will also attempt to demonstrate the effectiveness of your choice of mouthwash (with and/or without brushing and flossing). To complete this exercise, you must:

1. Understand the relationship between fermentation and the production of acid in saliva and on the teeth.
2. Understand the two procedures used in this exercise to estimate bacterial activity: direct counting and measurement of biological activity.
3. Understand the relationship between dietary sugar and the activity of the streptococci and lactobacilli in the oral cavity.

## MATERIALS NEEDED FOR THIS LABORATORY

For each saliva sample, the following are needed:

1. One tube of Snyder test agar for each saliva sample and one tube to be used as a control.
2. Three tubes of Rogosa SL agar (with about 20 ml/tube).
3. One tube with 9 ml of sterile saline.
4. Three sterile petri dishes.
5. One sterile beaker (50 ml) or sterile tube for the collection of saliva.
6. Sterile pipettes.
7. Small blocks of sterile paraffin for chewing to stimulate saliva production.

## LABORATORY PROCEDURE

For each saliva sample:

1. Obtain three sterile petri dishes and label them "Undiluted," "1:10," and "1:100."
2. Obtain three tubes of Rogosa SL agar (20 ml/tube), one tube of sterile saline (9 ml/tube), and one tube of Snyder test agar. Be sure the tubes of medium have been melted and cooled to about 48–50°C.
3. Chew the paraffin block to stimulate production of saliva. Collect at least 2.5 ml of saliva in the sterile beaker or tube provided for that purpose.
4. At this time, double-check to make sure the medium has been cooled to about 48–50°C.
5. Examine and follow the flow chart shown in Fig. E14-1.
6. Transfer 1 ml of saliva to one of the tubes of Rogosa agar. Mix well and pour into the plate labeled "Undiluted."

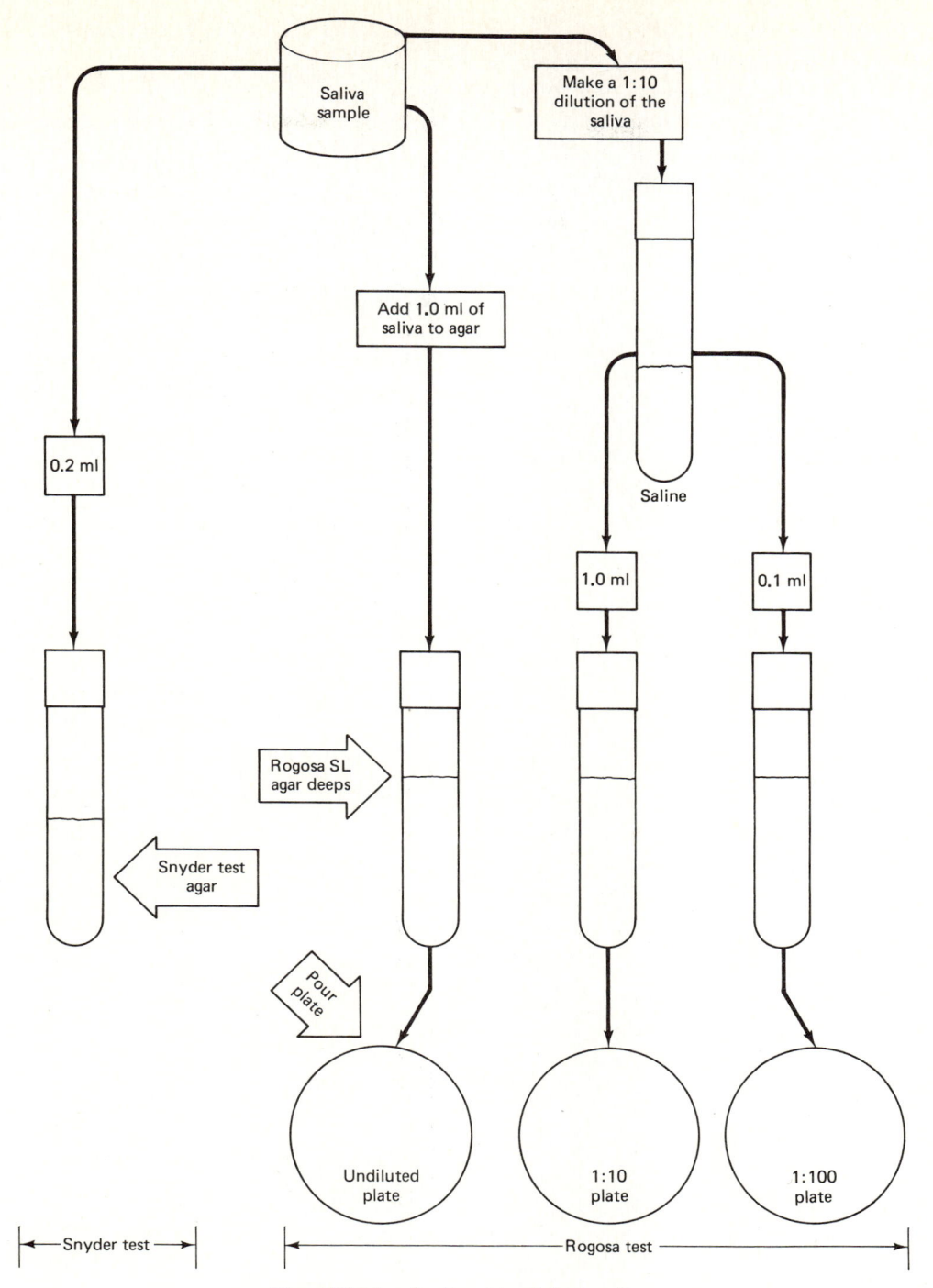

**Figure E14-1** Lactic-acid activity in saliva.

7. Transfer a second milliliter of the saliva to the tube of saline. Mix well and then transfer 1.0 ml of the saline dilution to one of the remaining tubes of medium, then transfer 0.1 ml to the other tube of medium. Mix well and pour into the respective plates ("1:10" and "1:100").

8. Finally, transfer 0.2 ml of saliva directly into the tube of Snyder test agar. Mix well and allow to solidify in the upright position.

9. Repeat the exercise as directed, after using mouthwash, flossing, brushing, or at another time of day.

10. Incubate all cultures for 72 hours.
11. Record the color of the Snyder test agar at 24-hour intervals, and count the colonies that develop on the Rogosa agar at 48 hours.
12. Complete the Laboratory Report Forms and answer all questions.

## REFERENCES

### Text References

1. Atlas, Ronald M., *Microbiology: Fundamentals and Applications.* Chapter 18.
2. Brock, Thomas D., David W. Smith, and Michael T. Madigan, *Biology of Microorganisms,* 4th ed. Chapter 14.
3. Nester, Eugene W., et al., *Microbiology,* 3rd ed. Chapter 22.
4. Stanier, R. Y., E. A. Adelberg, and J. Ingraham, *The Microbial World,* 4th ed. Chapter 28.
5. Stanier, R. Y., et al., *Introduction to the Microbial World.* Chapter 15.
6. Wistreich, George A., and Max D. Lechtman, *Microbiology,* 4th ed. Chapter 29.

### Resource References

1. Finegold, Sidney M., and W. J. Martin, *Bailey and Scott's Diagnostic Microbiology,* 6th ed. Chapters 14, 17, and 26.
2. Lennette, E. H., ed., *Manual of Clinical Microbiology,* 3rd ed. Chapter 4.

# LABORATORY PROTOCOL

## Exercise 14 — Lactobacillus Activity in Saliva

Check each step when you complete it.

### First Day

_____ 1. Obtain and label all materials as directed in the exercise:

> Three petri dishes
> One tube of Snyder test agar
> Three tubes of Rogosa SL agar
> One tube of saline
> Sterile pipettes

_____ 2. Collect at least 2.5 ml of saliva in the sterile beaker or tube. You may need to chew the paraffin to stimulate saliva production.

_____ 3. Review the flow diagram in Fig. E14-1.

_____ 4. Make a 1:10 dilution of the saliva by transferring 1.0 ml of saliva to the tube containing 9.0 ml of saline.

_____ 5. Add saliva and diluted saliva to the melted Rogosa SL agar, mix well, and pour into the appropriately labeled plates:

| Plate Labeled | Should Receive Rogosa Agar Seeded With |
|---|---|
| Undiluted | 1.0 ml of undiluted saliva |
| 1:10 | 1.0 ml of the 1:10 saline dilution |
| 1:100 | 0.1 ml of the 1:10 saline dilution |

_____ 6. Add 0.2 ml of saliva to the Snyder test agar, mix well, and allow to solidify.

_____ 7. Incubate all cultures for 72 hours.

### Second Day

_____ 8. Count the colonies on the Rogosa agar plates at 48 hours and again at 72 hours.

_____ 9. Record the color of the Snyder test agar at 24, 48, and 72 hours.

_____ 10. Record the results on the chart provided in the Laboratory Report Form and answer all the questions in the form.

# NOTES

Name: _____

## LABORATORY REPORT FORM

### Exercise 14—Lactobacillus Activity in Saliva

1. Complete the chart using the laboratory data collected from this exercise.

**BACTERIAL COUNT OF SALIVA AND TIME OF COLOR CHANGE**

|  | Before Brushing | After Brushing |
|---|---|---|
| Bacterial count in Rogosa SL agar in 48 hours. | _____ | _____ |
| Bacterial count in Rogosa SL agar in 72 hours. | _____ | _____ |
| Color of Snyder test agar in 24 hours. | _____ | _____ |
| Color of Snyder test agar in 48 hours. | _____ | _____ |
| Color of Snyder test agar in 72 hours. | _____ | _____ |

2. Complete the following chart. Obtain data from other groups in the laboratory.

| Student Number | Time of Color Change (in hours) | Number of Colonies (#/ml) | Number of Fillings or Cavities |
|---|---|---|---|
| 1. | _____ | _____ | _____ |
| 2. | _____ | _____ | _____ |
| 3. | _____ | _____ | _____ |
| 4. | _____ | _____ | _____ |
| 5. | _____ | _____ | _____ |
| 6. | _____ | _____ | _____ |
| 7. | _____ | _____ | _____ |
| 8. | _____ | _____ | _____ |
| 9. | _____ | _____ | _____ |
| 10. | _____ | _____ | _____ |

## ANSWER THE FOLLOWING QUESTIONS

1. What correlations, if any, can be observed in the data collected and displayed above?

2. List at least three factors that might affect the results of the Snyder and Rogosa test procedures.

   a.

   b.

   c.

3. How would the results change after the use of mouthwash?

4. How would the results change after eating candy?

5. How would you explain results showing that one of the students in your sample had a very low susceptibility to dental caries on the basis of the Snyder or Rogosa test, yet actually had a large number of dental caries and/or fillings?

# Exercise 15
# Antibiotic Sensitivity Testing: The Kirby-Bauer Procedure

Successful isolation and identification of clinically significant bacteria from a clinical specimen is only half of the task expected of the microbiology laboratory. Virtually every isolate that may have clinical significance must be tested to determine its antibiotic sensitivity pattern.

There are two reasons for this: First, the physician needs to have the sensitivity pattern available to protect the patient against the possibility that this particular isolate may be resistant to the commonly used antibiotics. Second, antibiotic sensitivity patterns have proven to be consistent for a given organism and can be used as an additional diagnostic characteristic.

## THEORETICAL BASIS OF THE KIRBY-BAUER PROCEDURE

The Kirby-Bauer procedure is a precisely controlled procedure developed to standardize antibiotic sensitivity testing. It relates the concentration of the antibiotic in the test medium to that used for therapeutic purposes. If the results of the Kirby-Bauer procedure are to have any validity at all, the components of the test must be rigidly controlled. These include:

**Standardized bacterial suspension.** The amount of antibiotic needed to inhibit a population of bacteria is proportionately related to the size of the population. The density of the bacterial suspension used in the Kirby-Bauer procedure is standardized by comparing it to a standardized suspension of known turbidity. The standard used is called a 0.5 MacFarland standard.

**Standardized concentrations of antibiotics in the disks.** All other things being the same, the diffusion rate of any chemical is directly proportional to the concentration of that chemical. The amount of antibiotic in the disk must be standardized if any consistency in the diameters of the zones is to be expected. When all other conditions are the same, a larger amount of antibiotic would produce a larger zone in the same amount of time. (How would you test the validity of this statement?)

**Standardized medium.** The Kirby-Bauer procedure uses a medium known as Muller-Hinton agar. This medium was chosen because its properties relative to diffusion of antibiotics and growth of bacteria are known. Any other medium would not necessarily allow the antibiotic to diffuse at the same rate and would not support the same amount of bacterial growth. If the medium is changed, the diameter of the zone of inhibition may change for a given concentration of antibiotic. The depth of the medium must also be standardized. If the medium is too deep, more of the antibiotic will diffuse downward and will be of slightly lower concentration, resulting in a smaller zone. The accepted depth of the medium is 4.0 mm.

**Standardized incubation time and temperature.** The rate of diffusion is temperature dependent and increases as the temperature is increased. Furthermore, if the time for diffusion is extended indefinitely, the concentration of the antibiotic in the medium will eventually be equal everywhere. Antibiotics are bacteriostatic, and if the concentration falls below a certain level (known as the *minimal inhibitory concen-*

*tration*), the bacteria may resume growth. An incubation temperature of 35 to 37°C for not more than 24 hours is required.

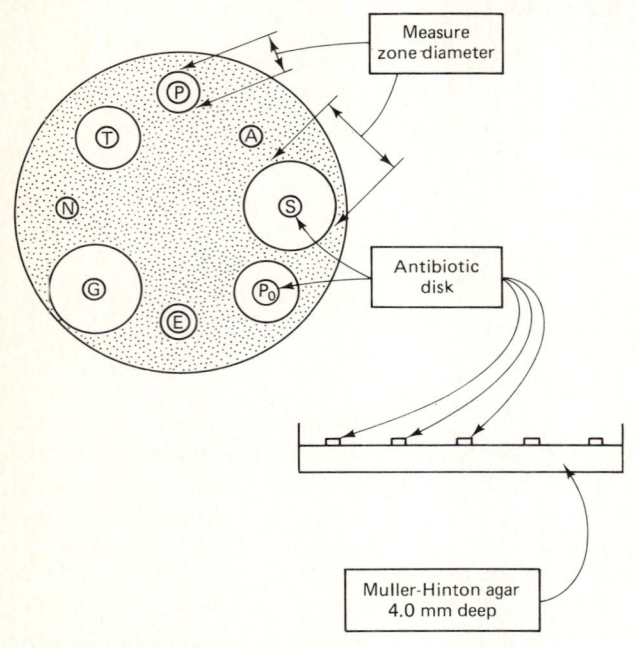

**Figure E15-1** Antibiotic sensitivity test (Kirby-Bauer).

The Kirby-Bauer procedure has proven to be a reliable method for determining antibiotic sensitivity of clinical isolates. The relationship between the laboratory findings and actual clinical use of antibiotics has proven to be similarly reliable when the standardized procedure is carefully followed.

## DETERMINATION OF SENSITIVITY

The Kirby-Bauer procedure uses three zone-diameter measurements for determining the resistance or sensitivity of the organism. Although the actual size of each zone will vary according to the antibiotic and the type of organism being tested, all antibiotics have published zone dimensions to indicate which of the following states apply to a given organism:

*Sensitive.* The minimum zone diameter, expressed in millimeters, that indicates the organism will be sensitive to the antibiotic at normal therapeutic doses. If a zone diameter larger than this value is observed, the organism is considered to be sensitive to the antibiotic.

*Resistant.* The minimum zone diameter, expressed in millimeters, that indicates the organism will be resistant to the antibiotic at normal therapeutic doses. If a zone diameter smaller than this value is observed, the organism is considered to be resistant to the antibiotic.

*Intermediate.* The range of zone diameter, expressed in millimeters, between sensitive and resistant zone diameters where individual differences between patients and bacterial strains may cause one particular combination to be resistant, while another combination might prove to be sensitive.

A typical Kirby-Bauer sensitivity chart will look similar to the following table. In the ex-

### ZONE DIAMETER SIGNIFICANCE

| Antibiotic | Disk Concentration | Resistant "R" | Intermediate "I" | Sensitive "S" |
|---|---|---|---|---|
| Penicillin | 10 units | <12 | 12-21 | >21 |
| Penicillin |  | *<21 | 21-28 | >28 |
| Ampicillin | 10 µg | <12 | 12-13 | >13 |
| Ampicillin |  | *<21 | 21-28 | >28 |
| Streptomycin | 10 µg | <12 | 12-14 | >14 |
| Polymyxin B | 300 units | <9 | 9-11 | >11 |
| Erythromycin | 15 µg | <14 | 14-17 | >17 |
| Gentamicin | 10 µg | <12 | 12-17 | >17 |
| Neomycin | 30 µg | <12 | 12-17 | >17 |
| Tetracycline | 30 µg | <14 | 14-19 | >19 |
| Chloramphenicol | 30 µg | <12 | 12-18 | >18 |
| Novobiocin | 30 µg | <17 | 17-21 | >21 |
| Kanamycin | 30 µg | <13 | 13-18 | >18 |

*Applies to *Staphylococcus aureus* only.

amples shown, the penicillin family of antibiotics requires a different interpretation when used with *S. aureus* than when used with other organisms. This is because the staphylococci have developed a high level of tolerance to the penicillins (why?). In the cases shown here, if the organism was *Staphylococcus aureus,* a minimum zone diameter of 28 mm would be required to interpret the isolate as being sensitive to ampicillin. Any other isolate would have to show a minimum zone diameter of 13 mm to be considered sensitive. A more extensive list of antibiotics and their Kirby-Bauer zone dimensions can be found in any of the Resource References.

## MINIMAL INHIBITORY CONCENTRATION

A more precise measure of sensitivity to antibiotics is possible using a procedure that determines the *minimal inhibitory concentration,* or *M.I.C.,* of an antibiotic. The M.I.C. is the minimum concentration of antibiotic that is needed to inhibit the growth of a given bacterial species under specified conditions. This value is of great importance because it indicates what concentration must be maintained in the body to successfully inhibit the growth of the bacterial pathogen. The M.I.C. is also used as a measure of resistance to the drug by a species of bacteria. What is the relationship between the M.I.C. and the diameter of the zone of inhibition produced by a given antibiotic?

Although the M.I.C. is not routinely determined in most laboratories, it may be requested when a bacterial infection does not respond to therapy. Consult the References for more information about the M.I.C. Exercise 16, dealing with the detection of mutant strains, contains a section on the isolation of antibiotic-resistant mutants that can be modified to demonstrate the M.I.C.

## LABORATORY OBJECTIVES

In this exercise, you will determine the antibiotic-sensitivity pattern for several of the bacteria you have been working with in previous exercises. You should determine the following:

1. Which, if any, of the antibiotics tested are effective against both gram-negative and gram-positive bacteria?
2. Are there any bacteria that appear to be particularly resistant to most antibiotics?
3. Are all the zones similar in appearance (not size)? Do some have hazy edges? Do some have colonies within the clear areas? If so, what is the significance of this observation?
4. Which antibiotic appears to be the most effective (at least against these bacteria)?
5. Which antibiotic appears to be the least effective (at least against these bacteria)?
6. What is the effect of antibiotic concentration on the diameter of the zone of inhibition?

## MATERIALS NEEDED FOR THIS LABORATORY

1. Eight 100-mm petri plates or four 150-mm petri plates containing Muller-Hinton agar, poured to a depth of 4 mm. You will need one 150-mm plate for each organism (using eight disks per plate) or two plates for each (using four disks per plate) if 100-mm plates are used.
2. Broth cultures of four of the organisms to be tested. You will do four of those listed below, and your partner will do the other four. The culture should be of the correct density.
3. Sterile swabs (at least one for each organism).
4. Antibiotic disks for eight different antibiotics. If available, disks with different concentrations of antibiotic may be used to demonstrate the effect of antibiotic concentration on zone size.
5. Disk dispensers or forceps.

## LABORATORY PROCEDURE

1. Obtain and label the required number of plates to perform the Kirby-Bauer antibiotic-sensitivity determination on four of the following bacteria (remember, your partner will do the other four):
    *Staphylococcus aureus*
    *Staphylococcus epidermidis*
    *Escherichia coli*
    *Klebsiella pneumoniae*
    *Pseudomonas aeruginosa*
    *Proteus vulgaris*

Demonstration Cultures
*Corynebacterium diphtheria*
*Salmonella typhimurium*
*Shigella sp.*
*Streptococcus pyogenes*

2. Note the color and initials coded on the disks. Confirm that the concentration of antibiotic on the disks agrees with those shown on the Kirby-Bauer chart.
3. Carefully swab the surface of each plate with the organism to be tested. Be particularly careful that the entire surface of the plate is covered so that a solid, continuous lawn of bacteria is obtained.
4. Place the antibiotic disks on the plates. Either use the dispensers the disks came in or apply them with lightly flamed forceps. Press lightly on each disk to be sure that it will adhere to the surface of the medium when the plate is inverted.
5. Repeat step 4 for each organism and for each different concentration of antibiotic. If you are testing the effect of antibiotic concentration, be sure to place those disks on the same plate (why?).
6. Incubate the plates for 24 hours.
7. Measure the zone of inhibition for each antibiotic with a centimeter ruler. Record the sensitivity pattern observed for each organism. Record the presence of any colonies in the cleared zone.

## REFERENCES

### Text References

1. Atlas, Ronald M., *Microbiology: Fundamentals and Applications.* Chapter 16.
2. Brock, Thomas D., David W. Smith, and Michael T. Madigan, *Biology of Microorganisms,* 4th ed. Chapter 7.
3. Jensen, Marcus, *Introduction to Medical Microbiology.* Chapter 9.
4. Nester, Eugene W., et al., *Microbiology,* 3rd ed. Chapter 10.
5. Stanier, R. Y., E. A. Adelberg, and J. Ingraham, *The Microbial World,* 4th ed. Chapters 3 and 31.
6. Stanier, R. Y., et al., *Introduction to the Microbial World.* Chapters 12 and 18.
7. Wistreich, George A., and Max D. Lechtman, *Microbiology,* 4th ed. Chapter 19.

### Resource References

1. Finegold, Sidney M., and W. J. Martin, *Bailey and Scott's Diagnostic Microbiology,* 6th ed. Chapter 36.
2. Gerhardt, P., ed., *Manual of Methods for General Bacteriology.* Pages 121, 229-234.
3. Lennette, E. H., ed., *Manual of Clinical Microbiology,* 3rd ed. Chapters 47 and 48. Appendixes 1, 2, and 3.
4. McGonagle, L. A., *Procedures for Diagnostic Microbiology.* Pages 1-12.

## LABORATORY PROTOCOL

### Exercise 15—Antibiotic Sensitivity Testing: The Kirby-Bauer Procedure

Check each step when you complete it.

**First Day**

_____ 1. Swab the required number of plates with 24-hour cultures of the test organisms. Remember that the density of the bacteria must be about 0.5 MacFarland turbidity.

_____ 2. Place antibiotic disks (four on 100-mm plates and eight on 150-mm plates) on the seeded medium. Be careful to space the disks evenly to avoid overlapping zones of inhibition.

_____ 3. Apply a small amount of pressure to each disk to ensure that it will adhere to the medium.

_____ 4. Repeat steps 1–3 for each organism and as needed for each concentration of antibiotic.

_____ 5. Incubate the plates for 24 hours at 37°C.

**Second Day**

_____ 6. Measure and record the diameters of any zones of inhibition observed. The measurements should be in millimeters and should include the diameters of the disks (6.0 mm).

   _____ a. Record zone size for each organism and antibiotic used in this experiment.

   _____ b. Indicate whether the organism is "R" (Resistant), "I" (Intermediate), or "S" (Sensitive).

   _____ c. Record the nature of the zone boundary. Is it hazy? Is it clear and sharp? What does this mean?

   _____ d. Record whether or not any colonies develop within the zone boundaries. What is the significance of any such colonies?

   _____ e. Attempt to determine (and plot) if a relationship exists between antibiotic concentration and zone size.

# NOTES

Name: _____

# LABORATORY REPORT FORM

**Exercise 15 — Antibiotic Sensitivity Testing: The Kirby-Bauer Procedure**

1. Complete the following charts. You may need to get some of the information from your laboratory partner or from other groups in the class.

### ANTIBIOTIC SENSITIVITIES OF SELECTED BACTERIAL SPECIES
(Record Zone Diameters)

| Antibiotic => | Pene. | Ampi. | Stre. | Poly. | Eryt. | Gent. | Neom. | Tetr. | Chlo. | Novo. | Kana. |
|---|---|---|---|---|---|---|---|---|---|---|---|
| P. aeruginosa | ___ | ___ | ___ | ___ | ___ | ___ | ___ | ___ | ___ | ___ | ___ |
| S. aureus | ___ | ___ | ___ | ___ | ___ | ___ | ___ | ___ | ___ | ___ | ___ |
| S. epidermidis | ___ | ___ | ___ | ___ | ___ | ___ | ___ | ___ | ___ | ___ | ___ |
| S. pyogenes | ___ | ___ | ___ | ___ | ___ | ___ | ___ | ___ | ___ | ___ | ___ |
| E. coli | ___ | ___ | ___ | ___ | ___ | ___ | ___ | ___ | ___ | ___ | ___ |
| P. vulgaris | ___ | ___ | ___ | ___ | ___ | ___ | ___ | ___ | ___ | ___ | ___ |
| K. pneumoniae | ___ | ___ | ___ | ___ | ___ | ___ | ___ | ___ | ___ | ___ | ___ |
| C. diphtheriae | ___ | ___ | ___ | ___ | ___ | ___ | ___ | ___ | ___ | ___ | ___ |

### ANTIBIOTIC SENSITIVITIES OF SELECTED BACTERIAL SPECIES
(Record as "R," "I," or "S")

| Antibiotic => | Pene. | Ampi. | Stre. | Poly. | Eryt. | Gent. | Neom. | Tetr. | Chlo. | Novo. | Kana. |
|---|---|---|---|---|---|---|---|---|---|---|---|
| P. aeruginosa | ___ | ___ | ___ | ___ | ___ | ___ | ___ | ___ | ___ | ___ | ___ |
| S. aureus | ___ | ___ | ___ | ___ | ___ | ___ | ___ | ___ | ___ | ___ | ___ |
| S. epidermidis | ___ | ___ | ___ | ___ | ___ | ___ | ___ | ___ | ___ | ___ | ___ |
| S. pyogenes | ___ | ___ | ___ | ___ | ___ | ___ | ___ | ___ | ___ | ___ | ___ |
| E. coli | ___ | ___ | ___ | ___ | ___ | ___ | ___ | ___ | ___ | ___ | ___ |
| P. vulgaris | ___ | ___ | ___ | ___ | ___ | ___ | ___ | ___ | ___ | ___ | ___ |
| K. pneumoniae | ___ | ___ | ___ | ___ | ___ | ___ | ___ | ___ | ___ | ___ | ___ |
| C. diphtheriae | ___ | ___ | ___ | ___ | ___ | ___ | ___ | ___ | ___ | ___ | ___ |

2. If you tested different concentrations of antibiotics, record the data here.

### RELATIONSHIP OF ANTIBIOTIC CONCENTRATION TO ZONE DIAMETER

Antibiotic: _____

| Antibiotic Concentration | Zone Diameter |
|---|---|
| ___ | ___ |
| ___ | ___ |
| ___ | ___ |
| ___ | ___ |
| ___ | ___ |

*RELATIONSHIP OF ANTIBIOTIC CONCENTRATION*

## ANSWER THE FOLLOWING QUESTIONS

1. Define *broad spectrum antibiotic*. Which, if any, of those tested in this exercise can be considered to be broad spectrum?

2. Explain the significance of the Kirby-Bauer reporting system. Why should you expect bacteria to be reported as resistant even when there is a clear and obvious zone of inhibition around the antibiotic disk?

3. How does this testing procedure relate to the M.I.C. testing procedure?

4. What is the significance of colonies that develop within otherwise clear zones of inhibition? If the laboratory report for one of your patients indicated colonies within the zone, what concerns would you have for your patient?

5. Why is *Pseudomonas aeruginosa* of such concern in burn patients and in immunologically compromised patients?

6. Why is it necessary to ensure that the zones of inhibition do not overlap? How would you demonstrate any synergistic effects of certain combinations of antibiotics?

# Exercise 16
# Detection of Mutant Strains of Bacteria

*Note:* Part Two of this exercise may be easily modified and used to determine the minimal inhibitory concentration (M.I.C.) by the agar plate method.

Mutations are changes in the base sequence of DNA that are transmitted to daughter cells and which may result in an inheritable change in the phenotype of a cell. If the new phenotype is selected against by the environment, the cell will disappear from the population. If, on the other hand, the change is selected for, the new cell type may gradually increase in number and eventually may become a significant part of the population. A "neutral" mutation—one that is neither selected for nor against—will have little, if any, effect on the distribution of the mutant in the population.

The development of a mutant population, then, is really the result of two processes: the mutation itself and the selection of that mutant by the environment. Let's look at an example. The ability to synthesize penicillinase was of no particular advantage to *Staphylococcus aureus* before penicillin became a widely used antibiotic. Penicillin did not become the well-known "wonder drug" until during World War II, when it was used extensively to treat wounds. After the war, it was made available to the public and was widely (indiscriminately?) used to treat virtually any bacterial infection. Some of you may remember when it was common to receive a penicillin shot when you went to a physician for a bad cold.

Nobody knows when the mutation for penicillin resistance first appeared in the population of *S. aureus*. Extensive selection for penicillinase-positive bacteria could not occur as long as penicillin was not widely distributed. As the antibiotic came into wide usage, the mutant (resistant) strain was suddenly placed at an advantage. It could inactivate the antibiotic, while the nonmutant (wild type) cells could not. Selection for the resistant members of the population was so extensive that it is now relatively uncommon to encounter an isolate of *S. aureus* that is sensitive to the original penicillin G.

Mutant strains cannot be detected unless we expose the bacteria to an environment that would select for that mutant. How would you detect penicillin-resistant bacteria unless you exposed the population to penicillin? While this seems self-evident, it does raise an interesting question: How can you prove that the mutation occurs independently from its selection? Consult your Text and Resource References for descriptions of the replica plate test and the fluctuation test.

Nevertheless, mutant individuals in a population can only be detected by exposing the population to an environment that selects for the mutant phenotype.

## EXPERIMENTAL DESIGN

In this exercise you will attempt to isolate mutant organisms from a population of bacteria. If a large number of cells are plated on a medium containing an antibiotic, then only those cells that are resistant to the antibiotic will grow. Similarly, in an experiment to isolate phage-resistant mutants, the bacteria may be mixed with a large number of viruses that normally lyse the cells. If any colonies develop when the phage-bacteria mixture is plated, we can assume they are resistant to the phage.

If we go one step further and count the number of bacteria we use in the experiment, we can then calculate, in a rough way, the relative number of mutants present in the original population. It is even relatively easy to estimate the mutagenic activity of, as an example, ultraviolet light, by comparing the number of mutant colonies that appear in a population of cells that was exposed to the ultraviolet light.

A typical experimental protocol might include the following steps:

1. Obtain the population of bacteria that you want to test. This could be a 24-hour culture of any bacterial species.
2. Divide the culture into two parts; expose one to a source of ultraviolet light that is known to produce mutants. Consult Exercise 20 to determine how long you will expose the cells to the UV light (you do not want to kill all the cells).
3. Complete a plate count on both these samples. Some of the bacteria will be killed by the UV light, so you must count both samples.
4. Plate both bacterial suspensions on the test medium. If you are testing for antibiotic resistance, you will add antibiotics to the medium. Consult Exercise 15 and the references given there to determine how much antibiotic to use (it must be larger than the M.I.C.). If you are testing for phage resistance, you should use a suspension of phage that will produce complete lysis (confluent plaques) of the sensitive cells. (Refer to Exercise 26.)
5. After the appropriate incubation period, count the colonies that grew on all plates. Do the appropriate calculations to determine the percent of cells that mutated.

$$\% \text{ mutant cells} = \frac{\text{Number of mutant cells}}{\text{Total number of cells}} \times 100$$

Bacteria vary greatly in their tendency to form mutant phenotypes. Some strains and/or species will yield very few mutants, while others will produce relatively large numbers of colonies on the antibiotic-containing medium. For best results, you and your partner should test different bacteria.

In this exercise, we will try to isolate three mutants of bacteria, each resistant to a different antibiotic. To accomplish this, you will add antibiotics to an agar medium, then inoculate the medium with sufficient bacteria to produce confluent growth. Your instructor will indicate which antibiotics to use. He or she may also have you test a bacterial culture that has been exposed to a mutagenic agent, and he or she may want you to attempt to determine the mutation rate.

In the second part of the exercise, you will try to demonstrate a concentration-dependent resistance to streptomycin. This part of the exercise can also be easily modified and used to determine the M.I.C. of an antibiotic (how?). Resistance to this antibiotic, unlike the penicillin family of antibiotics, often follows a concentration-dependent pattern. This means that the mutant cells will be resistant to certain amounts of streptomycin and not to others. For example, one mutant might be resistant to up to 40 $\mu$g/ml, while another might be resistant to up to 70 $\mu$g/ml. How do you think the results would change if penicillin were used for Part Two? Consult your references for some very interesting discussions about why this is so.

## LABORATORY OBJECTIVES

Bacterial mutations appear randomly in the population at a predictable rate. The mutation rate can be increased with certain mutagenic agents, but the resulting mutations are still randomly determined. A mutant population must be selected for by applying the correct environmental conditions. You should:

1. Understand that the mutations selected for in this exercise occurred independently of their selection and that almost any other type of mutant could have been detected by changing the selective environment.
2. Understand that the rate of mutations, but not their randomness, can be changed by mutagenic agents such as ultraviolet light.
3. Understand the relationship that exists between the indiscriminate use of antibiotics

and the development of resistant strains of pathogenic bacteria.

## MATERIALS NEEDED FOR THIS LABORATORY

1. Nutrient agar pours with 19 ml of medium. These deeps should be melted and cooled to 50°C. Ten tubes of medium will be needed.
2. Antibiotic stock solutions: The stock solutions should be prepared in such a way so that when 1 ml of the solution is added to the medium pours, the final concentration of antibiotic will exceed the minimal inhibitory concentration for the organism you are testing. You will need stock solutions of streptomycin and any three other antibiotics. Some examples are:

| Antibiotic | Stock Solution Concentration | Medium Concentration |
|---|---|---|
| Penicillin | 1000 µg/ml | 50.0 µg/ml* |
| Streptomycin | 1000 µg/ml | 50.0 µg/ml* |
| Ampicillin | 500 µg/ml | 25.0 µg/ml |
| Polymyxin B | 500 units/ml | 25.0 units/ml |
| Erythromycin | 1000 µg/ml | 50.0 µg/ml* |
| Gentamicin | 100 µg/ml | 5.0 µg/ml |
| Tetracycline | 250 µg/ml | 12.5 µg/ml* |
| Chloramphenicol | 500 µg/ml | 25.0 µg/ml |
| Kanamycin | 250 µg/ml | 12.5 µg/ml |

*These values based on estimated *in vivo* dosage. Other values based on published M.I.C. data.

3. A nutrient broth suspension of the organism you are testing. If you are supplied with agar-slant cultures, add 5.0 ml of nutrient broth to the agar-slant culture, then shake gently to suspend the bacteria. Your instructor will indicate whether or not you should expose the bacteria to a mutagen.
4. Optional supplies (as needed):
   - Plate-count supplies (Exercise 24)
     - Dilution blanks
     - Agar pours
     - Petri plates
   - Ultraviolet light (Exercise 20)
     - Plastic petri dishes
5. Bent glass spreaders. These will be used to spread the bacterial suspensions over the surface of the medium. They may be sterilized by dipping in ethanol (95%) and then burning the ethanol off with the Bunsen flame.

## LABORATORY ASSIGNMENT

Before beginning your laboratory assignment for this exercise, determine whether or not your instructor has any additional assignments for you. For example, he or she may want you to do a plate count on the bacterial suspension (Exercise 23) or he or she may want you to expose the suspension to ultraviolet light (Exercise 20).

### Part One: Concentration-Independent Resistance

1. Obtain four sterile petri dishes. Label three of them with the antibiotic, its concentration, and the name of the bacterial suspension to be tested. The fourth plate should be labeled "Control."
2. Obtain one tube of melted agar. Add 1.0 ml of an antibiotic stock solution to the tube, then pour the medium into the appropriately labeled petri dish and allow it to solidify.
3. Repeat step 2 with two more antibiotic solutions.
4. The control is prepared by adding 1.0 ml of sterile water to a tube of medium, mixing, and then pouring into the plate labeled "Control."
5. When you sterilize the bent glass spreaders in the next step, be sure that you do not hold the rods over anything flammable. Ignited alcohol may drop off the rod and set fire to whatever it lands on. Clear the countertop of all unnecessary materials.
   *Note:* If you are not attempting to determine mutation rate, you may use a cotton swab to cover the surface of the medium with bacteria (as you did in Exercise 15).
6. Transfer 0.1 ml of the bacterial suspension to each of the agar plates. Quickly spread the suspension over the surface with the sterilized bent glass spreaders. (See Fig. E16-1.)

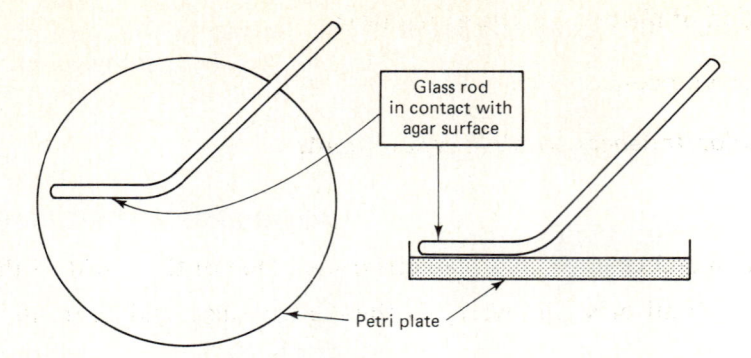

**Figure E16-1** Spread-plate inoculation with bent glass rod.

7. Incubate all plates for at least 48 hours.
8. Count the colonies that develop and complete any tests (as directed by the instructor) to verify that the colonies are antibiotic-resistant mutants of your test organism.

## Part Two: Concentration-Dependent Resistance

1. Obtain six sterile petri dishes. Label five of them with the concentration of antibiotic and the name of the bacterial suspension. Label the sixth plate "Control."
2. Obtain six tubes of melted medium. Keep the tubes in a water bath to ensure that they do not solidify. Add the required amounts of the streptomycin stock solution to the tubes, mix by rocking, and pour into the appropriately labeled petri dishes. Use the following table.

| Tube or Plate No. | Use ml of Stock Solution | Concentration in Medium |
|---|---|---|
| 1 | 0.2 ml | 10 µg/ml |
| 2 | 0.4 ml | 20 µg/ml |
| 3 | 0.6 ml | 30 µg/ml |
| 4 | 0.8 ml | 40 µg/ml |
| 5 | 1.0 ml | 50 µg/ml |
| 6 (Control) | 1.0 ml of water | Control |

*Note:* If the concentration of streptomycin in the medium were a critical part of this exercise, you would need to be sure that exactly 1.0 ml of additional volume (i.e., 0.2 ml of antibiotic stock solution + 0.8 ml of sterile water) was added to each tube. Why?

3. Transfer 0.1 ml of the bacterial suspension to each of the agar plates. Quickly spread the suspension over the surface of the medium with a sterilized bent glass spreader. Use the same precautions as in Part One.
4. Incubate all plates for at least 48 hours.
5. Count the colonies that develop and record the results on the Laboratory Report Form.

## REFERENCES

### Text References

1. Atlas, Ronald M., *Microbiology: Fundamentals and Applications*. Chapter 8.
2. Brock, Thomas D., David W. Smith, and Michael T. Madigan, *Biology of Microorganisms*, 4th ed. Chapter 9.
3. Jensen, Marcus, *Introduction to Medical Microbiology*. Chapter 9.
4. Nester, Eugene W., et al., *Microbiology*, 3rd ed. Chapter 8.
5. Stanier, R. Y., E. A. Adelberg, and J. Ingraham, *The Microbial World*, 4th ed. Chapter 14.
6. Stanier, R. Y., et al., *Introduction to the Microbial World*. Chapters 8 and 18.
7. Wistreich, George A., and Max D. Lechtman, *Microbiology*, 4th ed. Chapters 8 and 19.

### Resource References

1. Gerhardt, P., ed., *Manual of Methods for General Bacteriology*. Chapter 13.
2. Lennette, E. H., ed., *Manual of Clinical Microbiology*, 3rd ed. Chapters 41 through 48.

# LABORATORY PROTOCOL

## Exercise 16—Detection of Mutant Strains of Bacteria

Check each step when you complete it.

### Part One: Concentration-Independent Resistant Mutants

**First Day**

_____ 1. Obtain and label four sterile petri dishes. One of these will be the control.

_____ 2. Add 1.0 ml of sterile water to the control, mix, and pour into the appropriate petri dish.

_____ 3. Add 1.0 ml of the stock antibiotic solutions to the remaining tubes, mix, and then pour into the appropriately labeled petri dishes.

_____ 4. While the plates are solidifying, prepare your bacterial suspension.

_____ 5. After the agar has solidified, transfer 0.1 ml of the bacterial suspension to each of the plates. Spread the fluid with a sterilized bent glass spreader.

_____ 6. Incubate all plates for at least 24 hours.

**Second Day**

_____ 7. Count all colonies that develop. Record your results on the Laboratory Report Form for this exercise.

### Part Two: Concentration-Dependent Resistant Mutants (Minimal Inhibitory Concentration)

**First Day**

_____ 1. Obtain and label six sterile petri dishes. One of these will be your control.

_____ 2. Obtain six tubes of melted medium. Use a water bath to keep them liquid.

_____ 3. Add 1.0 ml of sterile water to one of the tubes, mix, and pour into the plate labeled "Control."

_____ 4. Add the requisite amounts of streptomycin to each of the remaining five tubes (see the table in the Laboratory Assignment section of this exercise), mix, and pour into the appropriately labeled plates.

_____ 5. While the medium is solidifying, prepare your bacterial suspension. Transfer 5 ml of sterile broth to a slant; mix well to suspend the bacteria.

_____ 6. Transfer 0.1 ml of the bacterial suspension to each of the petri plates, then spread the suspension with a sterilized bent glass spreader.

_____ 7. Incubate all plates for at least 24 hours.

**Second Day**

_____ 8. Count all the colonies that develop. Record all data on the Laboratory Report Form.

# NOTES

Name: _____

# LABORATORY REPORT FORM

## Exercise 16—Detection of Mutant Strains of Bacteria

1. Complete the following table.

**NUMBER OF MUTANT COLONIES**

| Antibiotic | Concentration in Medium | Number of Colonies |
|---|---|---|
| _____ | _____ | _____ |
| _____ | _____ | _____ |
| _____ | _____ | _____ |

2. Carefully study each colony to verify that it is morphologically similar to your test organism. Complete any additional tests necessary to confirm that you have not picked up contaminants.

3. Complete the following table.

**NUMBER OF MUTANT COLONIES**

ANTIBIOTIC: _____

| Tube Number | Concentration in Medium | Number of Colonies |
|---|---|---|
| 1 | _____ | _____ |
| 2 | _____ | _____ |
| 3 | _____ | _____ |
| 4 | _____ | _____ |
| 5 | _____ | _____ |
| 6 (Control) | _____ | _____ |

4. Once again, check the colonial morphology to verify that the colonies are not contaminants.

## ANSWER THE FOLLOWING QUESTIONS

1. Describe how the mutant colonies were selected for in this exercise.

2. How can similar selection occur in the real world?

3. How would you modify this experiment to detect phage-resistant mutants?

4. How would the results of this experiment change if penicillin were used?

5. How would you modify this experiment to determine the minimal inhibitory concentration (M.I.C.) of the antibiotics given in the table on page 165?

6. List three mutagenic agents that would increase the number of mutant colonies.

   a.

   b.

   c.

7. Discuss one experiment that would demonstrate that the actual mutations occurred independently of their selection.

8. What are R factors; how were they first detected? Why are they examples of the selection of mutants?

# Exercise 17
# Serological Reactions

Serological reactions are specific reactions involving antigens and antibodies. The term *serological* is used because serum is the most convenient source of antibody. Antibody-antigen reactions are typically very specific, often approaching the specificity of enzymatic reactions. It is because of this high level of specificity that serological reactions can be used for diagnostic work.

When one of the components of the reaction is known (either the antigen or the antibody), the other component can be detected with a very high level of sensitivity and reliability. If a reaction occurs when antibody is added to an unknown mixture, it is virtually certain that the antigen was present. When the term *homologous* is used to describe an antibody or an antigen, it indicates that the antigen and antibody can react with each other. That is, an antigen will react with its homologous antibody, and, barring cross reactions, only with its homologous antibody. Similarly, an antibody will react only with its homologous antigen.

**THE USE OF SEROLOGICAL REACTIONS**

Serological reactions may be used to detect the presence of either antibody or antigen, depending upon the test being performed. For example, if you want to determine if a patient has been exposed to a certain virus (or bacteria, or rickettsia), you could mix the patient's serum with a known sample of antigen; if a reaction occurs, the antibody is present in your patient's serum. Conversely, an unknown virus can be tested against a known sample of antibodies; if reaction occurs, the identity of the virus will be known.

There are several types of *in vitro* serological reactions that are commonly used in diagnostic laboratories. Some are very much more sensitive than others, but the more sensitive tests are sometimes more difficult to run. In this exercise you will study the type of serological reaction known as *agglutination*.

Agglutination occurs when a particulate antigen, such as a cell, reacts with its homologous antibody and agglutinates, or appears to form clumps. Characteristically, the antigen is *polyvalent*, while the antibody is *divalent*. When the two are mixed, increasingly large complexes of antigen-antibody are formed that appear to clump together. The agglutination complexes form because the divalent antibody molecule reacts with two particulate antigens. Each antigen, because it is polyvalent, can also react with several antibody molecules. The repeating complex

<Antigen>−<Antibody>−<Antigen>−

−<Antibody>−<Antigen>

eventually gets so large that the particles appear to clump together.

One of the most commonly used agglutination reactions is *blood typing* (A, B, O, and Rh+ or D) of blood. In this exercise you will determine the blood type of other students in the class. You will also use agglutination reactions to *serotype* bacterial unknowns. (What does serotype mean?) The *febrile agglutination test* will be used to demonstrate how known antigen (killed bacterial cells) will be used to detect antibodies in a patient's serum. In all types of serological reactions, however, the antibody and antigen must react. These reactions are highly specific, and it is this specificity that makes the reactions so important diagnostically.

## LABORATORY OBJECTIVES

Serological reactions are characteristically very specific and, therefore, of diagnostic value. In this exercise, you should:

1. Appreciate the diagnostic value of serological reactions.
2. Understand why a known antigen *or* antibody can be used to identify an unknown sample of virus or serum.
3. Begin to understand the basis of skin tests and the nature of the substance used in such testing.
4. Understand the nature of agglutination reactions.

## MATERIALS NEEDED FOR THIS LABORATORY

### For Blood Typing and Typing

1. Anti-A, anti-B, and anti-Rh typing sera.
2. Glass slides. If available, depression slides (with three depressions) or embossed slides should be used. (See Fig. E17-1.)

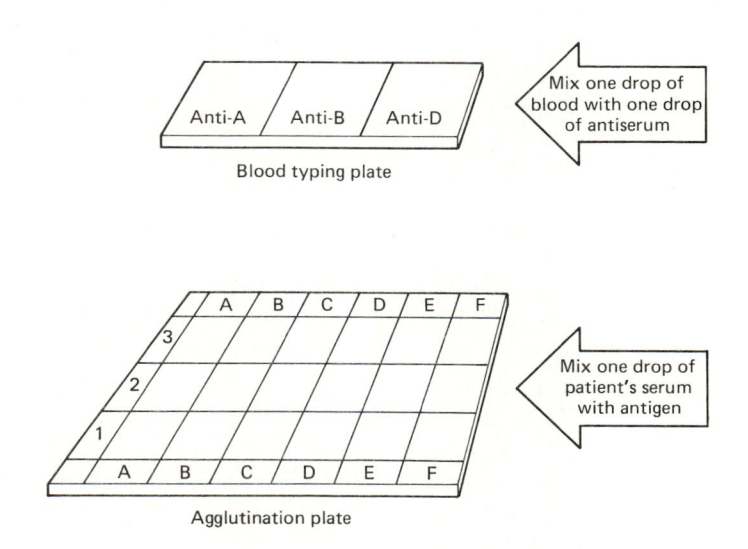

**Figure E17-1** Slide-agglutination plates.

3. Lancets and alcohol (70%) swabs or wipes.
4. Toothpicks or applicator sticks.

### For Serotyping of Unknown Bacteria

1. Polyvalent *Salmonella* O Antiserum Set A-1 (Difco No. 2892-32: *Salmonella* Poly A-1) and Positive Control Antigen (Difco No. 2840-56: *Salmonella* O Antigen, Group B).
2. Phenolized saline suspensions of *Salmonella*. If you are instructed to prepare your own suspensions, you will need a 24-hour agar-slant culture of the organism and at least 1.0 ml of phenolized saline (0.5% saline in 0.85% saline).
3. Glass slides or depression slides.

### For the Febrile Agglutination Test

1. Febrile agglutination sets (i.e., Difco No. 2407-32-7), including positive and negative controls.
2. Unknown serum samples (may be obtained from local hospitals or clinics).
3. Agglutination plates or slides.

## LABORATORY ASSIGNMENT AND PROCEDURES

*Note:* You should complete all parts of this exercise unless told otherwise by your laboratory instructor. Each type of serological reaction is presented independently of each other.

### Slide Agglutination Reactions— Blood Typing

A patient's blood type may be determined by mixing the patient's (the unknown) red blood cells with serum known to agglutinate either Group A, Group B, or Group D (Rh+) cells. Blood typing is almost always done when a person enters a hospital. It is particularly important when there is a possibility of surgery, or when there may be questions about the blood type of a newborn infant.

### Procedure

1. Obtain a glass slide and draw three circles with a wax pencil. Each circle should be about the size of a nickel. If depression slides are used, draw a circle around each of three depressions.
2. Obtain samples of the three antisera (anti-A, anti-B, and anti-D). Have them ready to use.
3. Obtain a lancet; open it and position it so that the tip does not touch the laboratory bench. (Leave the point *in the package* until you are ready to use it.)
4. Carefully clean the tip of your finger with the alcohol swab or wipe. Allow the cleansed area to air dry. (Do not recontaminate it by drying it with a paper towel or cloth.)
5. Sometimes it is easier to get a good sample of blood if you knead or milk the fingertip. Often, soaking your fingertip in warm water or simply allowing the hand to hang below the core of the body for a few minutes will produce good blood flow.
6. Lance the fingertip by quickly stabbing it with the lancet. A quick jab is usually least painful and most effective.
7. Allow one drop of blood to fall into each of the circles on the slide. Pinch a sterile swab against the lancet wound until a clot forms and blood flow stops (about one minute). Dispose of the swab in a container provided for that purpose.
8. *Drop* one drop of the antisera into the appropriate circles. *Do not allow the dropper to touch the blood.* Drop the antisera into the blood from about 1 cm above the slide.
9. Using separate toothpicks for each of the three mixtures, carefully mix the blood and antiserum to obtain a homogeneous mixture. Agglutination should occur (if it is going to) almost immediately.
10. Record the type of blood on the Laboratory Report Form.

| Agglutination With | Indicates |
| --- | --- |
| Anti-A | Type A |
| Anti-B | Type B |
| Anti-A and Anti-B | Type AB |
| Neither Anti-A nor B | Type O |
| Anti-D (Anti-Rh) | Type D (Rh+) |

Failure to agglutinate with the Rh antiserum indicates Rh− (type d).

### Slide Agglutination Reactions— Serological Identification of Bacterial Unknowns

Bacterial cells have two major types of antigens on their cell surface. Both types are species, and sometimes strain, specific. These antigens have proven to be very valuable in epidemiological studies when it is important to learn if all cases of a disease are caused by the same strain of bacteria.

These antigens are found either on the bacterial flagella or in the outer membrane of gram-negative cells. The *flagellar antigens,* or *H-antigens,* can be removed when the flagella are broken off the cell. The antigens found in the outer membrane, sometimes called *somatic* or *O-antigens,* remain bound to the cells even after relatively vigorous treatment. When a laboratory wishes to serotype an unknown, the technician will usually determine the serotypes of both the O and H antigens present on the bacterial cell surface.

In this exercise you will use a polyvalent antibody that reacts with almost all strains of *Salmonella.* This polyvalent antibody is really a mixture of antibodies and is used to confirm that the bacterial isolate actually is *Salmonella*. If a positive reaction occurs with the polyvalent antibody, then the unknown is tested with increasingly more specific antibodies until its complete serotype is known.

### Procedure

1. If you are not supplied with a suspension of bacteria in phenolized saline, you may prepare your own suspensions. Place about 1 ml of the phenolized saline into a clean test tube. Transfer a loopful of the unknown bacterial culture (from an agar slant) to the phenolized saline and mix well.
2. Draw three circles on a clean glass slide with a wax pencil. The circles should be about the size of a nickel. If depression slides are used, draw a circle around each depression.
3. Each of the circles will contain different combinations of test and control mixtures. Usually one drop is a sufficient amount of reagent. As before, *drop* the test and control reagents onto the slide—do not allow droppers to touch the slide or the reagents.
4. Prepare the test mixtures according to the following table:

| Test Circle: | One Drop of: | + | One Drop of: |
| --- | --- | --- | --- |
| First circle: | Phenolized saline | + | Unknown suspension of bacteria |
| Second circle: | Test antiserum | + | Unknown suspension of bacteria |
| Third circle: | Test antiserum | + | Suspension of the known positive control |

5. Mix the reagents in each of the circles with a toothpick. Use a different toothpick for each mixture. Agglutination should occur almost immediately.
6. The first mixture, with saline and the unknown bacterial suspension, should not agglutinate (negative control). The third mixture, with the antiserum and the suspension of known positive bacteria, should agglutinate (positive control). The second mixture will agglutinate *only if* the organism has antigens that are homologous to one of the antibodies in the polyvalent mixture on its flagella or outer membrane.
7. Record results on the Laboratory Report Form for this exercise.

### FEBRILE AGGLUTINATION TEST

The febrile agglutination test uses known antigens to determine if a patient's serum contains antibody to one or more of the antigens contained in the set. The antigens are bacterial cells that have been killed and stained. In this test, the antigen is the known and you will test for the presence of homologous antibody in an unknown (a patient's serum). Known positive controls should always be run with the test. A typical febrile agglutination test set (i.e., Difco #2407-32-7) includes the following cellular antigens and controls:

*Proteus* OX19 antigen
*Brucella abortus* antigen
*Salmonella* O, Group D (Typhoid O) antigen
*Salmonella* H Group a (Paratyphoid A) antigen
*Salmonella* H Group b (Paratyphoid B) antigen
*Salmonella* H Group d (Typhoid H) antigen
Positive control serum
Negative control serum

Agglutination with any of the antigens in the set indicates that the patient has been exposed to the organisms and may recently have had, or now has, the indicated disease.

This test is usually run by mixing one drop of the patient's serum with one drop of each of the test antigens. If agglutination occurs, and all the controls show the expected reactions, the patient is presumed to have been exposed to the antigen (how else would he/she have the antibody?). Additional testing is usually necessary to determine if the patient is harboring the infectious agent and to obtain more specific information about which strain of antigen he/she was exposed to.

### FEBRILE AGGLUTINATION TEST
#### Indications of Positive Reactions

| Agglutination With | → Possible Pathology |
| --- | --- |
| *Proteus* OX-19 | Rocky Mountain spotted fever or typhus |
| *Brucella abortus* | Brucellosis |
| *Salmonella* O (D) | Typhoid fever |
| *Salmonella* H (a) | Paratyphoid fever |
| *Salmonella* H (b) | Paratyphoid fever |
| *Salmonella* H (d) | Typhoid fever |

### Procedure — Qualitative
### Rapid Slide Test

1. There are six bacterial antigens in the set, and each one of them must be tested against the unknown serum and against the two controls.
2. Depending upon the type of plates or slides available, mark three rows of six test areas (these may be depressions or embossed areas on a single, large glass or porcelain plate). One row will be used for the unknown serum and the other two will be used for the controls. See Fig. E17-1.
3. Place one drop of serum into each depression or test area in the appropriate row.
4. As before, *drop* one drop of antigen into the drop of serum. Do not allow the dropper tip to touch the plate or the drop of serum.
5. Mix each with a separate toothpick or applicator stick.
6. Agglutination should occur almost immediately; the results should be observed in about one minute.
7. Record your observations on the Laboratory Report Form for this exercise.
8. Your laboratory instructor may want to demonstrate how to obtain quantitative results with this test procedure.

## REFERENCES

### Text References

1. Atlas, Ronald M., *Microbiology: Fundamentals and Applications.* Chapters 14 and 16.
2. Brock, Thomas D., David W. Smith, and Michael T. Madigan, *Biology of Microorganisms,* 4th ed. Chapter 16.
3. Jensen, Marcus, *Introduction to Medical Microbiology.* Chapter 13.
4. Nester, Eugene W., et al., *Microbiology,* 3rd ed. Chapter 21.
5. Wistreich, George A., and Max D. Lechtman, *Microbiology,* 4th ed. Chapter 22.

### Resource References

1. Buchanan, R. E., and N. E. Gibbons, eds., *Bergey's Manual of Determinative Bacteriology,* 8th ed. Pages 298–319.
2. Finegold, Sidney M., and W. J. Martin, *Bailey and Scott's Diagnostic Microbiology,* 6th ed. Chapters 37 and 39.
3. Gerhardt, P., ed., *Manual of Methods for General Bacteriology.* Chapter 20.
4. Lennette, E. H., ed., *Manual of Clinical Microbiology,* 3rd ed. Chapters 16 and 51.

# LABORATORY PROTOCOL

## Exercise 17—Serological Reactions

Check each step when you complete it.

### First Day

_____ 1. Determine the blood type of your laboratory partner. Record your results on the Laboratory Report Form. Record the results obtained for the entire class in the appropriate table.

_____ 2. Complete one serological identification of a culture of *Salmonella*. Be sure to include a test control in your experiment. Record your results on the Laboratory Report Form.

_____ 3. Complete a febrile agglutination test on one serum sample. Include controls in your procedure. Record your results on the Laboratory Report Form.

_____ 4. Answer all questions on the Laboratory Report Form.

# NOTES

Name: _____

## LABORATORY REPORT FORM

**Exercise 17—Serological Reactions**

1. Record your blood type:

    Blood type: (A, B, AB, O) _____

    Blood type: (D or Rh) _____

2. Determine the blood type distribution for your class. Obtain the data from the other students in the class:

    **NUMBER OF STUDENTS IN CLASS:** _____

    | Blood Type | Number | Percent |
    |---|---|---|
    | Type A | _____ | _____ |
    | Type B | _____ | _____ |
    | Type AB | _____ | _____ |
    | Type O | _____ | _____ |
    | | | |
    | Type Rh+ (Type D) | _____ | _____ |
    | Type Rh− (Type d) | _____ | _____ |

3. Describe the agglutination reactions observed in the slide agglutination serotyping of the bacterial cultures.

4. Fill in the following chart using your data from the febrile agglutination test.

    | Antigen | Postive or Negative |
    |---|---|
    | *Proteus* OX-19 | _____ |
    | *Brucella abortus* | _____ |
    | *Salmonella* O (D) | _____ |
    | *Salmonella* H (a) | _____ |
    | *Salmonella* H (b) | _____ |
    | *Salmonella* H (d) | _____ |

## ANSWER THE FOLLOWING QUESTIONS

1. What type of antibodies would be found in the serum of a person with Group B red blood cells?

2. What is meant by the term *universal donor*?

3. Why is blood group O, type Rh− used in replacement transfusions in newborn infants?

4. Distinguish between O and H antigens.

5. Explain the word *febrile*.

6. Explain the word *agglutination*.

# Exercise 18
# Assay of Antimicrobial Agents: Disk-Diffusion Methods

How effective are the antiseptics and disinfectants you will be using when you are engaged in your chosen careers? Does silver nitrate really work when used to rinse the eyes of the newborn? Can you effectively sterilize the counter or tabletop? What is the best agent for soaking oral thermometers? Can the antiseptic used to cleanse the skin also be used to sterilize instruments? These are questions you will need to be concerned with throughout your career.

There are three ways to evaluate the effectiveness of antimicrobial agents. Each of the techniques has certain advantages and certain disadvantages; all are useful in the proper circumstances. These techniques are:

**Disk-diffusion technique.** This technique is the same as the one used to assay antibiotics (the Kirby-Bauer procedure) and has the same advantages and disadvantages. In the testing of antimicrobial agents, it is important to remember that this method cannot distinguish between *bacteriocidal* and *bacteriostatic* agents, since no effort is made to remove the agent from contact with the bacteria.

**Use-dilution technique.** In this procedure, the bacteria are removed from the inhibitory agent. It is possible to determine if the bacteria were killed (not just inhibited) by contact with the agent. This technique can, therefore, distinguish between bacteriocidal and bacteriostatic agents. It is demonstrated in the next laboratory exercise (Exercise 19).

**Phenol coefficient.** The phenol coefficient is the ratio of the concentration of inhibitory agent to the concentration of phenol that will kill the bacteria being tested in ten minutes, but not in five minutes. The procedure is a quantitative comparison of a bacteriocidal agent with a standard—in this case, phenol. This procedure will tell you how effective the agent is relative to phenol. We will not determine a phenol coefficient, although it may be demonstrated for you or it may be assigned as an optional exercise.

## THEORETICAL AND PRACTICAL CONSIDERATIONS

The single common factor shared by all inhibitory agents is that they must react with some cellular component and cause changes that make that component inactive. Most inhibitory agents owe their effectiveness to their ability to react with proteins in such a way as to denature them or to their ability to affect the integrity and function of membranes. *Heavy metals,* such as the ones that will be tested in this exercise, react with proteins, usually the sulfhydryl groups, disrupting their tertiary structure (denaturing them). *Detergents* are effective because they alter the surface tension of water and disrupt many structures that rely on hydrophobic properties for their integrity or their function. *Disinfectants* usually combine the denaturing activity of metals or halides with the wetting activity of detergents. The commercial product *Betadyne* is an example of the combination of a halide (iodine) and a detergent. In every case, the agent must react with the cell if it is to be effective.

Several factors must be weighed when

choosing the correct antimicrobial agent. One such factor is relative toxicity of the agent and its likelihood of being absorbed across the skin or mucous membranes. For example, different agents and/or different concentrations of the same agent are used for cleansing the skin than are used for the mucous membranes (e.g., mouthwash). Another factor to be considered is how the agent affects the surface that is to be sanitized. Plastic laminates, ceramics, stainless steel, enameled surfaces, and wood are examples of surfaces that may react adversely when certain agents are used. They may etch, become discolored, or may pose additional problems because of their porosity. Some of these problems will be addressed in Exercise 19.

You must also consider the nature of the fluid the bacteria are suspended in. It is almost always true that bacteria live in fluids containing compounds that will also react with the inhibitory agents. The problem posed is this: If the agent reacts with components of the medium, it can no longer react with the cell, and a higher concentration of the inhibitory agent will be needed. For example, the amount of inhibitor needed to suppress the growth of bacteria suspended in saline will be very different (much lower) than the amount needed to kill the same number of bacteria suspended in a protein solution.

It is possible to determine what effect the suspension components will have by performing these tests using different suspending fluids. What does it mean if a markedly smaller inhibitory zone is produced when the medium is changed but the concentration of agent on the disk and the density of the bacterial suspension are left constant?

## LABORATORY OBJECTIVES

In this exercise, you will determine the relative effectiveness of three heavy metal ions, phenol, formaldehyde, and three antimicrobial agents of your choice. The disk-diffusion technique will be used to compare these agents on two different types of media. You should:

1. Understand why this procedure is useful for measuring bacteriostatic activity and not bacteriocidal activity.
2. Understand why the components of the suspending medium might interfere with the antimicrobial activity of certain inhibitory agents.
3. Gain some appreciation for the choice of which agent to use in a given circumstance.

## MATERIALS NEEDED FOR THIS LABORATORY

1. Solutions of agents to be tested:

| Mercuric chloride | 0.1 M |
| Silver nitrate | 0.2 M |
| Zinc sulfate | 0.2 M |
| Phenol | 10% |
| Formaldehyde | 5% |
| Disinfectants | Three of your choice, prepared beforehand and made up to double the concentration recommended for use. |
| Antiseptics | Three of your choice, prepared beforehand and made up to double the concentration recommended for use. |

2. Filter paper disks, 6.0 mm in diameter, sterile.
    *Note:* Unless instructed otherwise, you should complete this exercise with one organism. Your partner will perform the assay on one other.
3. The choice of media should include one that contains components that react with the inhibitory agents. For example, nutrient agar will give different zone diameters than thioglycolate medium.
4. Sufficient plates of at least two types of medium so that the heavy metals, phenol, and formaldehyde can be tested using a single disk on each plate, and so the disinfectants and antiseptics can be tested with not more than four disks/plates. If you use the agents suggested in number 1 (above), you will need at least seven plates of each type of medium for each organism being tested.
5. Sterile swabs.
6. Forceps.
7. Twenty-four-hour broth cultures of:
    *Escherichia coli*
    *Staphylococcus aureus*
    *Bacillus subtilis*

## LABORATORY PROCEDURE

1. Obtain the required number of plates of each medium and label them according to organism and inhibitory agent.

    Media:   Nutrient agar
             Thioglycolate medium
    Inhibitory agents:  Mercuric chloride
                        Silver nitrate
                        Zinc sulfate
                        Phenol
                        Formaldehyde
                        Up to six commercial products of your choice

2. Swab each plate with 24-hour cultures of the test organisms, one organism per plate. Be sure to completely cover the medium surface to produce confluent growth with a solid lawn of bacteria.

3. Prepare disks by dipping *only the edge* of each disk into the solutions of inhibitory agents. Allow the disks to absorb the solution by capillary action. Then touch the disk to paper toweling to remove any excess fluid. This is essential to prevent excess fluid from running over the surface of the medium and distorting the zone.

4. Place the disks on the surface of the seeded media and apply a small amount of pressure to ensure adherence to the medium when the plates are inverted during incubation. Remember that the heavy metals, formaldehyde, and phenol are tested with one disk per plate, but that the commercial products may have up to four disks per plate.

5. Incubate all plates for at least 24 hours.

6. Measure the diameters of the zones of inhibition for each agent against each bacterium. Compare the results obtained on each medium. Construct a chart of the results from the entire class so that the zones for all media, all agents, and each bacterium can be compared.

## REFERENCES

### Text References

1. Atlas, Ronald M., *Microbiology: Fundamentals and Applications*. Chapter 10.
2. Brock, Thomas D., David W. Smith, and Michael T. Madigan, *Biology of Microorganisms*, 4th ed. Chapter 7.
3. Jensen, Marcus, *Introduction to Medical Microbiology*. Chapter 8.
4. Nester, Eugene W., et al., *Microbiology*, 3rd ed. Chapter 5.
5. Wistreich, George A., and Max D. Lechtman, *Microbiology*, 4th ed. Chapter 17.

## LABORATORY PROTOCOL

### Exercise 18—Assay of Antimicrobial Agents: Disk-Diffusion Methods

Check each step when you complete it.

**First Day**

*Note:* Unless instructed otherwise, you should complete this exercise with one organism. Your partner will perform the assay on one of the others.

_____ 1. Swab the required number of plates with 24-hour cultures of the test organisms. You will need at least seven plates of each medium for each bacterium tested.

_____ 2. Prepare and apply the disks. Remember to be careful that there is no excess fluid on the disks. Use one disk per plate for the heavy metals, formaldehyde, and phenol, but you may use up to four disks per plate with the commercial products.

_____ 3. Incubate all plates for at least 24 hours.

**Second Day**

_____ 4. Measure and record all zone diameters.

_____ 5. Compare the zones of inhibition for each organism and inhibitory agent. Also compare the results obtained on each medium. Explain any differences that may be observed.

_____ 6. Make notes about any changes that are observed in the medium within the zones of inhibition. Is there any evidence of precipitation or other reactions?

# NOTES

Name: _____

## LABORATORY REPORT FORM

**Exercise 18—Assay of Antimicrobial Agents: Disk-Diffusion Methods**

1. Complete the chart on the following page. You will need to get the information from your laboratory partner or from other groups in the class.

## ANSWER THE FOLLOWING QUESTIONS

1. Explain any differences that might be observed in the diameters of the zones of inhibition observed.

2. Define *bacteriocidal* and *bacteriostatic*.

3. How would you modify this experiment to determine if these inhibitory agents are bacteriocidal?

4. Using the reference material in your texts, determine what class of inhibitory agent each of the disinfectants and antiseptics are and list their chemical mode of action. (For example, household bleach is a halide derivative, and its germicidal effects are due to oxidation of cell components.)

EXERCISE 18

## DIAMETERS OF ZONES OF INHIBITION
## FOR SELECTED ANTIMICROBIAL AGENTS

*Name of organism:*

|  | *Nutrient Broth* | *Thioglycolate* | *Nutrient Broth* | *Thioglycolate* |
|---|---|---|---|---|
| Mercuric chloride | | | | |
| Silver nitrate | | | | |
| Zinc sulfate | | | | |
| Formaldehyde | | | | |
| Phenol | | | | |
| Disinfectants: | | | | |
| 1. | | | | |
| 2. | | | | |
| 3. | | | | |
| Antiseptics: | | | | |
| 1. | | | | |
| 2. | | | | |
| 3. | | | | |

# Exercise 19
# Assay of Antimicrobial Agents: Use-Dilution Methods

Bacteriostatic agents are effective only if they remain in contact with the bacteria. A good model to use to explain this phenomenon is one of a reversible reaction where dissociation occurs if the inhibitory agents are removed:

$$\text{Bacterial Cell + Inhibitory Agent} \longleftrightarrow \text{Inhibited Cell}$$

Bacteriocidal agents, on the other hand, irreversibly damage the cell so that the cells cannot grow even when the agent is removed from contact with the cell. The bacteria are dead. A model that can be used to explain this phenomenon is that of an irreversible reaction:

$$\text{Bacterial Cell + Inhibitory Agent} \rightarrow \text{Inhibited Cell}$$

The experimental procedure necessary to prove bacteriocidal activity requires that the agent be removed from contact with the cells. This can be accomplished by washing the cells in saline (centrifuging and rinsing) or by diluting the agent down to a concentration that is not effective. If the cells have been irreversibly damaged, they will not grow when inoculated into fresh, agent-free medium.

All inhibitory agents, bacteriocidal or bacteriostatic, must react with the cells if they are to be effective. In the last exercise we saw how components of the medium can interfere with the action of inhibitors by reacting with them before they can react with the cells. This ability to react with inhibitors may make it possible for some media to appear to reverse the effects of certain inhibitors. Components of the medium may leach the agent from the cells and chemically bind it. When the agent is removed, the cell is able to grow normally. This phenomenon will occur when the inhibitory agent has a stronger propensity to react with compounds in the medium than with the cells.

Another factor affecting the usefulness of agents is their ability to soak into or to penetrate pores in the surface being sanitized. Wood, skin, unsealed concrete and mortar, and some unglazed ceramics have pores large enough for bacteria to settle in. (Why are surgical rooms tiled?) If the agents cannot penetrate the pores, they will be unable to sanitize the surface. The combination of detergents with other inhibitors, typically halides (e.g., Betadyne), enhances the effectiveness of the disinfectant by changing the surface tension and improving on its penetration qualities. It is also possible to increase the exposure time by leaving the disinfectant on the surface for longer periods. This is not always practical.

## EXPERIMENTAL DESIGN

In this experiment, toothpicks and stainless steel pins will be soaked in a bacterial suspension for several minutes and then transferred to solutions of the same inhibitory agents used in Exercise 18. The stainless steel pins are comparable to nonporous surfaces, while the toothpicks are comparable to porous surfaces.

When the pins and toothpicks are transferred to tubes of broth, any residual living bacteria will grow. The volume of broth in the tubes (about 10 ml) is so large relative to the amount of agent adhering to the pins and toothpicks that they will be diluted out to a concentration that is not effective (even if it is a bacteriostatic agent). If no growth is observed, the bacteria can be assumed to have been killed and the agent said to be bacteriocidal. Bacteriostatic agents will not have killed the bacteria and growth will occur in the broth tubes. An agent that produced a zone of inhibition in Exercise 18 but did not kill the cells in this exercise is clearly bacteriostatic, whereas one that did both would be bacteriocidal. (What if it did neither?)

## LABORATORY OBJECTIVES

In this exercise, you will determine if certain inhibitory agents are bacteriocidal. The inhibitory agents that were tested in the last exercise will be tested by a use-dilution procedure that removes contaminated objects from contact with the antimicrobial agents. You should:

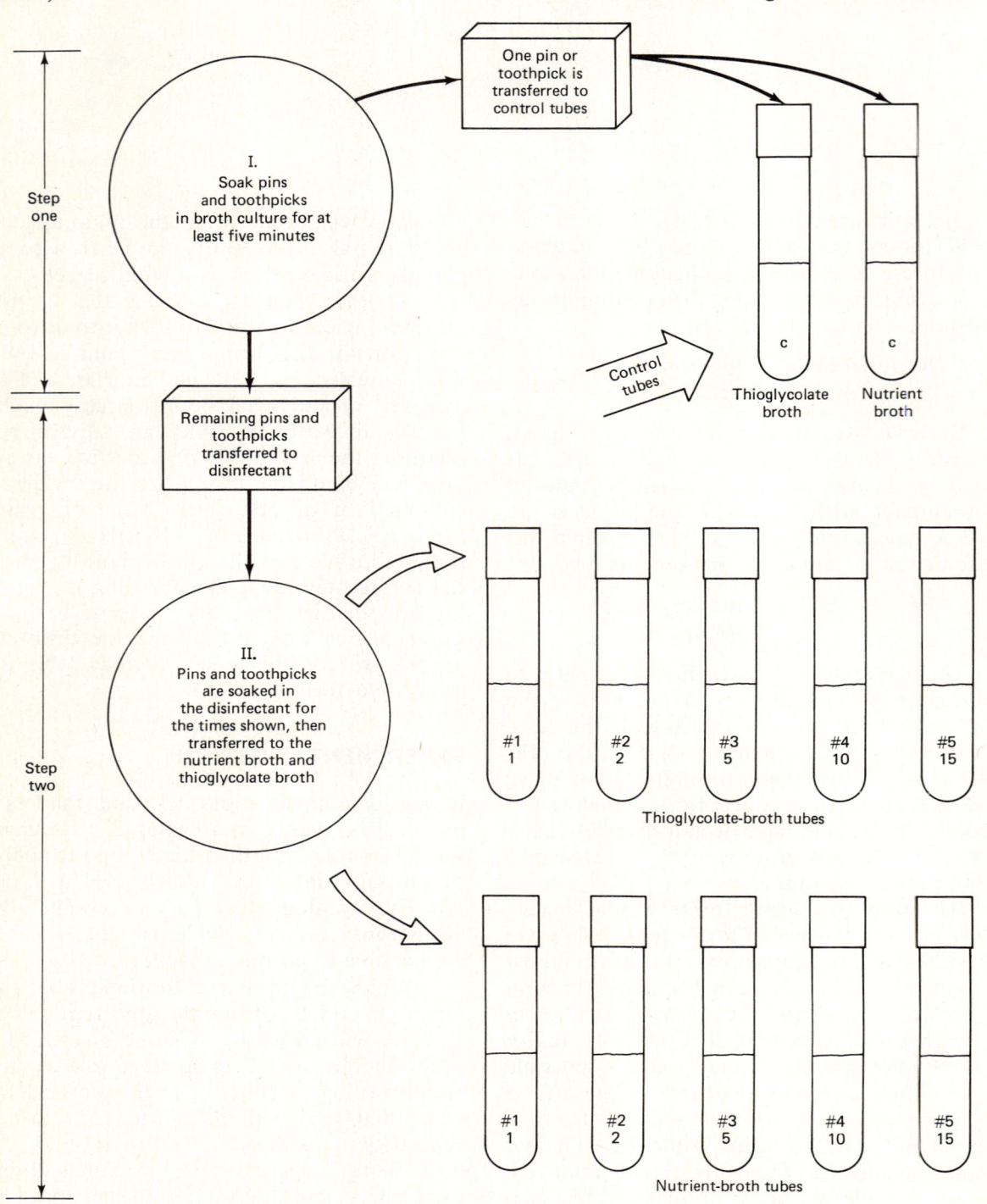

**Figure E19-1** Use-dilution assay of antimicrobial agents and disinfectants.

1. Understand the difference between bacteriostatic and bacteriocidal agents.
2. Explain how to test for the two types of microbial agents.
3. Explain the difference between antiseptic and disinfectant.
4. Explain the difference between *sanitize* and *sterilize*.

**MATERIALS NEEDED FOR THIS LABORATORY**

Each student will need the following items:

1. One of the solutions of inhibitory agents used in the previous exercise (Exercise 18).
2. Two empty, sterile petri dishes.
3. Eleven toothpicks or eleven stainless steel pins.
4. Twenty-four-hour broth culture of one of the following:
    *Escherichia coli*
    *Staphylococcus aureus*
    *Bacillus cereus*
5. Six tubes of sterile nutrient broth and six tubes of sterile thioglycolate broth. There should be about 10 ml of medium in each tube.
6. Forceps.

**LABORATORY PROCEDURE**

*Note:* You will do half of this exercise and your laboratory partner will do the other half of the exercise with you. One of you will complete the exercise with pins and the other will use the toothpicks. Your group will complete the experiment with one of the bacterial species, while another group in the class will complete the exercise with another one. The instructions are the same for either the pins or the toothpicks.

1. Label both sets of tubes according to the times for soaking in the inhibitory agents:
    No. 0 = Control, zero time
    No. 1 = One-minute soak
    No. 2 = Two-minute soak
    No. 5 = Five-minute soak
    No. 10 = Ten-minute soak
    No. 15 = Fifteen-minute soak
    You should have two sets of tubes: one of nutrient broth and one of thioglycolate broth.
2. Pour 10 ml of the bacterial suspension into one of the petri dishes and 10 ml of the inhibitory agent solution into the other petri dish.
3. Place eleven toothpicks (or pins) into the bacterial suspension and allow them to soak for about five minutes. Be sure that the toothpicks (or pins) are covered by the broth suspension.
    *Note:* When making the transfer in the next step, and in all others in this exercise, you should *drop* the pin or toothpick into the broth. Do not allow the forceps to touch the broth in the tubes. Flame the forceps between each use.
4. Transfer one pin to the "0" tube of nutrient broth, and one pin to the "0" tube of thioglycolate broth. This will be your "zero time" or control tube—it will show that the bacteria are viable and that they can be transferred on the pins or toothpicks.
    *Note:* After placing all but one of the pins (or toothpicks) into the inhibitory agent, you will transfer them at timed intervals, one at a time, into either the nutrient broth or the thioglycolate. The timing will begin when you first transfer the pins from the bacterial suspension to the solution of inhibitory agent.
5. Transfer all the remaining ten pins (or toothpicks) to the petri plate containing the inhibitory agent. Remember to drop the pins (or toothpicks) into the broth. Begin timing immediately.
6. According to the times given below, transfer one pin to the nutrient broth and one pin to the thioglycolate broth. The times are:
    One minute → Tube No. 1
    Two minutes → Tube No. 2
    Five minutes → Tube No. 5
    Ten minutes → Tube No. 10
    Fifteen minutes → Tube No. 15
7. Incubate all tubes for 24 hours.
8. Carefully dispose of the petri dishes by placing them in a pan of disinfectant provided for that purpose.
9. Record whether or not growth occurred in each tube of medium. Compare results with both media and prepare a chart of the re-

sults from the data obtained by the rest of the class.

## REFERENCES
### Text References

1. Atlas, Ronald M., *Microbiology: Fundamentals and Applications.* Chapter 10.
2. Brock, Thomas D., David W. Smith, and Michael T. Madigan, *Biology of Microorganisms,* 4th ed. Chapter 7.
3. Jensen, Marcus, *Introduction to Medical Microbiology.* Chapters 8 and 9.
4. Nester, Eugene W., et al., *Microbiology,* 3rd ed. Chapter 5.
5. Wistreich, George A., and Max D. Lechtman, *Microbiology,* 4th ed. Chapter 17.

# LABORATORY PROTOCOL

## Exercise 19—Assay of Antimicrobial Agents: Use-Dilution Methods

Check each step when you complete it.

### First Day

*Note:* You will do half of this exercise and your laboratory partner will do the other half of the exercise with you. One of you will complete the exercise with pins and the other will use the toothpicks. Your group will complete the experiment with one of the bacterial species, while another group in the class will complete the exercise with another one. The instructions are the same for either the pins or the toothpicks.

_____ 1. Label six tubes of nutrient broth and six tubes of thioglycolate broth according to the instructions given in Laboratory Procedure, step 1.

_____ 2. Pour the bacterial suspension and the inhibitory agent solution into separate sterile petri dishes.

_____ 3. Place eleven toothpicks or eleven stainless steel pins into the petri plate with the bacterial suspension. Be sure that the picks (or pins) are completely covered with the broth culture.

*Note:* When making the transfer in the next step, and in all others in this exercise, you should *drop* the pin or toothpick into the broth. Do not allow the forceps to touch the broth in the tubes. Flame the forceps between each use.

_____ 4. Five minutes later, transfer one pin (or pick) to each of the tubes of broth labeled "0."

*Note:* After placing all but one of the pins (or toothpicks) into the inhibitory agent, you will transfer them at timed intervals, one at a time, into either the nutrient broth or the thioglycolate. The timing will begin when you first transfer the pins from the bacterial suspension to the solution of inhibitory agent.

_____ 5. Immediately transfer the remaining pins to the plate with the solution of inhibitory agent. Begin timing immediately.

_____ 6. At each of the times given in the exercise (1, 2, 5, 10, and 15 minutes) transfer one pin (or one toothpick) to the appropriately labeled tube of nutrient broth and thioglycolate broth.

_____ 7. Incubate all tubes for 24 hours.

### Second Day

_____ 8. Score each tube for growth or no growth as a function of time of soaking in the inhibitory agent.

_____ 9. Determine if the growth medium had any effect on the activity of the agent used (did it take longer to kill the bacteria in one of the media?).

_____ 10. Compare the results of this exercise with those obtained in Exercise 18. Which of the agents, if any, are bacteriostatic and which are bacteriocidal?

# NOTES

Name: _____

## LABORATORY REPORT FORM

**Exercise 19—Assay of Antimicrobial Agents: Use-Dilution Methods**

1. Complete the charts that follow. You will need to get the information from your laboratory partner or from other groups in the class.

## ANSWER THE FOLLOWING QUESTIONS

1. Explain any differences that might be observed between the killing times noted for the pins and toothpicks.

2. Define *antiseptic, disinfectant, sanitize,* and *sterilize.*

3. How would you modify this experiment to determine the phenol coefficients of the agents used?

4. Using the reference material in your texts, determine what class of inhibitory agent each of the disinfectants and antiseptics are and list their chemical mode of action. (For example, household bleach is a halide derivative, and its germicidal effects are due to oxidation of cell components.)

## TIME NEEDED TO KILL BACTERIA ON NONPOROUS SURFACES (STEEL PINS)

| Name of Organism: | | | | |
|---|---|---|---|---|
| | *Nutrient Broth* | *Thioglycolate* | *Nutrient Broth* | *Thioglycolate* |
| Mercuric chloride | | | | |
| Silver nitrate | | | | |
| Zinc sulfate | | | | |
| Formaldehyde | | | | |
| Phenol | | | | |
| Disinfectants: | | | | |
| 1. | | | | |
| 2. | | | | |
| 3. | | | | |
| Antiseptics: | | | | |
| 1. | | | | |
| 2. | | | | |
| 3. | | | | |

## TIME NEEDED TO KILL BACTERIA ON POROUS SURFACES (TOOTHPICKS)

| Name of Organism: | | | | |
|---|---|---|---|---|
| | *Nutrient Broth* | *Thioglycolate* | *Nutrient Broth* | *Thioglycolate* |
| Mercuric chloride | | | | |
| Silver nitrate | | | | |
| Zinc sulfate | | | | |
| Formaldehyde | | | | |
| Phenol | | | | |
| Disinfectants: | | | | |
| 1. | | | | |
| 2. | | | | |
| 3. | | | | |
| Antiseptics: | | | | |
| 1. | | | | |
| 2. | | | | |
| 3. | | | | |

# Exercise 20
# Ultraviolet Light as a Bacteriocidal Agent

*A word of caution:* Ultraviolet light can damage the retina of the human eye. *Under no circumstances should you look into the lamp used in this experiment. Avoid having reflected light shine into your eyes.*

Ultraviolet light is often used to sterilize surfaces of objects that cannot be conveniently sterilized by other methods. It is also used to sterilize air entering and leaving certain types of "clean rooms." Its effectiveness as a bacteriocidal or germicidal agent is based on the light-absorbing properties of nucleic acids.

## THEORETICAL BASIS FOR BACTERIOCIDAL EFFECTS OF ULTRAVIOLET LIGHT

Nucleic acids strongly absorb ultraviolet light when the wavelength of the light falls between 250 and 260 nm (nanometers). Absorbance falls off rapidly as the wavelength is either increased or decreased from these values. The effectiveness of ultraviolet light as a bacteriocidal agent is directly proportional to its absorbance by the nucleic acids of living cells.

When the light (a form of energy) is absorbed, the nucleic acids are raised to a higher energy level. In this elevated energy state they show a markedly greater tendency to react chemically, producing intramolecular bonds (usually thiamine dimers) as well as some reactions with compounds found in the cytoplasmic fluids. Unless these reactions are corrected, mutations will occur. When a large number of mutations occur, or when certain critical genes are mutated (lethal mutations), the cell will be unable to divide and the bacteriocidal effects of ultraviolet radiation will have been demonstrated.

Ultraviolet light is, therefore, both a *mutagenic agent* (causes mutations in living cells) and, if applied in a large enough dose, a *bacteriocidal agent*. It has proven to be a valuable tool for both purposes and is frequently used to increase mutation rate in molecular genetics experiments as well as being used as a sterilizing agent in many other applications. Perhaps the most commonly observed germicidal application of ultraviolet radiation is the exposure of hair clippers to ultraviolet light in barber shops and beauty salons. It is quite effective in sterilizing surfaces and air if it is used properly.

## PRACTICAL CONSIDERATIONS

Ultraviolet light is absorbed by many other biological compounds, especially proteins. It is absorbed by dead cells as well as living cells, and by any nucleic acids or nucleotides that may be present in the medium in which the bacteria are suspended. When these other ultraviolet-absorbing materials are present, they effectively protect the living cells from damage by "shading" them from the ultraviolet radiation. If ultraviolet light is to be effectively used as a bacteriocidal agent, it must be absorbed by the nucleic acids of living cells. Any other factor that absorbs ultraviolet light will reduce its apparent effectiveness. Unfortunately, any fluid or particle that is likely to have bacteria in or on it will also contain other ultraviolet-absorbing substances. The effectiveness of ultraviolet light may not be as good in the real world as it is under laboratory conditions.

When choosing a germicidal lamp, it is

essential that it emit the correct wavelength of light of sufficient intensity. If it is to be effective as a germicide, it must emit a radiation between 250 and 260 nm that is of sufficient intensity to kill the cells. There are many ultraviolet lights on the market; not all of them emit a wavelength of light that is germicidal.

## LABORATORY OBJECTIVES

In this exercise you will determine the time needed to kill suspensions of bacteria exposed to ultraviolet radiation. To complete this exercise, you will need to:

1. Understand the mechanism by which ultraviolet radiation accomplishes its bacteriocidal effects.
2. Understand the reasons why certain components of the medium may protect suspended bacteria from ultraviolet-induced lethal effects.
3. Understand the need for specified wavelengths and be able to explain the difference between ultraviolet lamps, black lights, and true germicidal lamps.
4. Explain the correct applications of ultraviolet-light sterilization procedures and practices.

## EXPERIMENTAL DESIGN

This experiment is designed to demonstrate the germicidal properties of ultraviolet light and that certain components of the suspending medium may protect bacteria from the ultraviolet radiation. To accomplish this, you will suspend bacteria in two different solutions: saline and nutrient broth. At certain time intervals, a loopful of the irradiated suspensions will be transferred to fresh nutrient agar. If any bacteria have survived, they will of course produce visible growth after appropriate incubation.

Since saline does not contain any ultraviolet-absorbing material, the bacteria should be killed rapidly. The other suspending medium (nutrient broth) contains ultraviolet-absorbing components and will show some protective effects. It should take longer to kill the bacteria in nutrient broth than it does in saline.

An alternate experiment to the one described here would be one where you exposed the same suspensions of bacteria (e.g., all in saline) to different ultraviolet-emitting devices that produced different wavelengths of light and then measured killing as a function of wavelength.

## MATERIALS NEEDED FOR THIS LABORATORY

1. Four petri dishes of nutrient agar. Divide the plates into four quadrants by marking them with a felt marker or wax pencil.
2. A 24-hour broth culture of *Escherichia coli*.
3. One tube each of sterile saline and nutrient broth.
4. Two empty, sterile petri dishes. If available, use the small (60-mm diameter) size.
5. Sterile pipettes.
6. An ultraviolet lamp that emits with a wavelength between 250 and 260 nm.

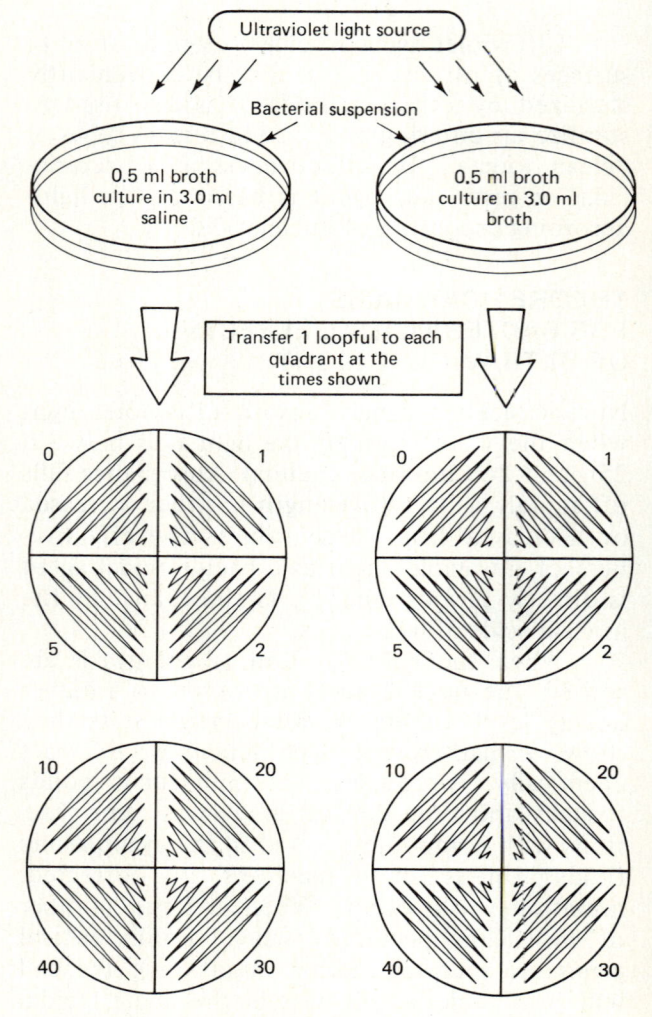

**Figure E20-1** Bacteriocidal effects of ultraviolet light.

## LABORATORY PROCEDURE

1. Obtain four petri dishes with nutrient agar. Divide each plate into four quadrants by marking the plates on the back with a wax pencil or felt pen.
2. Label two of the plates "Saline" and the other two "Nutrient Broth." Label the quadrants on the first saline plate "0," "1," "2," "3"; and "5," "10," "15," and "30" on the second saline plate. Label the nutrient broth plates in a similar manner.
3. Transfer 1 ml of the broth culture to the tube of saline and 1 ml to the tube of nutrient broth. Mix well.
4. Pour about 3 ml of the mixtures into separate, empty petri dishes.
5. Transfer one loopful of each bacterial suspension in the petri dishes to the quadrants of the plates labeled "0." These are the "zero time" controls; they are needed to show that the bacteria were viable before being placed under the ultraviolet light source.

   *A word of caution:* Ultraviolet light can damage the retina of the human eye. *Under no circumstances should you look into the lamp used in this experiment. Avoid having reflected light shine into your eyes.*
6. Place the *open* (the dishes must be open because the ultraviolet light may be absorbed by the plastic or glass cover) petri dishes with the two bacterial suspensions under an ultraviolet light source. Be sure that the entire plate is exposed to the light. Begin timing immediately after putting the plates under the light source.

   *Note:* The distance between the lamp and the petri dishes is important because it determines the amount of radiation striking the surface of the fluid. The lamp should be positioned in its holder (if one is provided) or not more than 15 inches from the surface of the fluid.
7. At the times given below, transfer one loopful of each suspension to the appropriate quadrant of the petri dishes containing the nutrient agar. The saline suspension will be transferred to the plates labeled "Saline," and the nutrient broth suspensions will be transferred to the plates labeled accordingly.

   | Exposure Time | Quadrant Number |
   | --- | --- |
   | 1 minute | "1" |
   | 2 minutes | "2" |
   | 3 minutes | "3" |
   | 5 minutes | "5" |
   | 10 minutes | "10" |
   | 15 minutes | "15" |
   | 30 minutes | "30" |

8. Incubate all plates for at least 24 hours.
9. Observe the plates for growth. Complete the Laboratory Report Forms for this exercise.

## REFERENCES

### Text References

1. Atlas, Ronald M., *Microbiology: Fundamentals and Applications.* Chapter 10.
2. Brock, Thomas D., David W. Smith, and Michael T. Madigan, *Biology of Microorganisms,* 4th ed. Chapters 8 and 9.
3. Nester, Eugene W., et al., *Microbiology,* 3rd ed. Chapter 5.
4. Stanier, R. Y., E. A. Adelberg, and J. Ingraham, *The Microbial World,* 4th ed. Chapter 13.
5. Stanier, R. Y., et al., *Introduction to the Microbial World.* Chapter 8.
6. Wistreich, George A., and Max D. Lechtman, *Microbiology,* 4th ed. Chapter 18.

### Resource References

1. Gerhardt, P., ed., *Manual of Methods for General Bacteriology.* Chapter 13.

# LABORATORY PROTOCOL

## Exercise 20—Ultraviolet Light as a Bacteriocidal Agent

Check each step when you complete it.

**First Day**

_____ 1. Turn on the ultraviolet light and allow it to warm up for a few minutes while you are getting everything else ready.

*A word of caution:* Ultraviolet light can damage the retina of the human eye. *Under no circumstances should you look into the lamp used in this experiment. Avoid having reflected light shine into your eyes.*

_____ 2. Obtain and label four petri dishes according to the instructions given in the laboratory exercise. You should have two plates labeled "Saline" and two labeled "Nutrient Broth."

_____ 3. The two sets of plates should be divided into four quadrants each and labeled "0," "1," "2," "3," "5," "15," "20," and "30."

_____ 4. Transfer 1 ml of the 24-hour broth culture to the tube of saline, and 1 ml to the tube of nutrient broth. (The reason for this step is to ensure that the numbers of bacteria in both parts of the experiment are the same, at least at the beginning of the experiment.)

_____ 5. Transfer 5 ml of the newly prepared suspensions (saline and nutrient broth) to empty, sterile petri dishes. If available, use the smaller (60-mm diameter) size.

_____ 6. Transfer one loopful from each to the quadrants of the plates labeled "0."

*Note:* The distance between the lamp and the petri dishes is important because it determines the amount of radiation striking the surface of the fluid. The lamp should be positioned in its holder (if one is provided) or not more than 15 inches from the surface of the fluid.

_____ 7. Place the petri dishes with the saline and nutrient broth suspensions under an ultraviolet light source. Be sure the dishes are open (why?) and be sure they are completely exposed to the ultraviolet light. Begin timing of the exposure to ultraviolet light as soon as the plates are placed under the light.

_____ 8. At the times given, transfer one loopful of each suspension to the appropriately labeled quadrants of the petri dishes:

| Exposure Time | Transfer to | Quadrant Number |
|---|---|---|
| 1 minute | → | "1" |
| 2 minutes | → | "2" |
| 3 minutes | → | "3" |
| 5 minutes | → | "5" |
| 15 minutes | → | "15" |
| 20 minutes | → | "20" |
| 30 minutes | → | "30" |

_____ 9. Incubate all plates for at least 24 hours.

**Second Day**

_____ 10. Record whether or not growth appeared in each of the quadrants.

_____ 11. Complete the Laboratory Report Form for this exercise.

# NOTES

Name: _____

## LABORATORY REPORT FORM

### Exercise 20—Ultraviolet Light as a Bacteriocidal Agent

1. Complete the chart using the laboratory data collected from this exercise.

**TIME NECESSARY TO KILL *Escherichia coli*
BY EXPOSURE TO ULTRAVIOLET LIGHT**

| Exposure Time | In Saline | In Nutrient Broth |
|---|---|---|
| 0 minutes | _____ | _____ |
| 1 minute | _____ | _____ |
| 2 minutes | _____ | _____ |
| 3 minutes | _____ | _____ |
| 5 minutes | _____ | _____ |
| 15 minutes | _____ | _____ |
| 20 minutes | _____ | _____ |
| 30 minutes | _____ | _____ |

## ANSWER THE FOLLOWING QUESTIONS

1. Explain any differences in the killing times observed.

2. Why must the petri dishes be left open during the exposure to ultraviolet light?

3. How would you modify this experiment to determine the mutagenic activity of ultraviolet light?

4. What important safety precaution must be observed when using ultraviolet light?

5. What is the most prominent source of ultraviolet radiation in the world? What health problem is associated with exposure to this source?

6. How would the results of this experiment change if the suspensions were twice as turbid as they were? What would happen if you doubled the intensity of the ultraviolet radiation?

7. Explain the difference between ionizing and nonionizing radiation.

8. Why is the distance between the light source and the surface of the fluid an important variable in this exercise? What would happen if you decreased the light intensity?

# Exercise 21
# Calibration of the Microscope

If your microscope is to be used to obtain quantitative data, it is necessary that it be calibrated so that accurate linear measurements can be made. In this exercise we will determine the diameter of the field of view and calibrate an ocular micrometer. In Exercises 22 and 23 we will use the calibrated microscopes to determine the sizes of red blood cells, yeast, and bacteria, and for cell counting.

## A REMINDER

The *stage micrometer* is a specially manufactured glass slide that has a precise linear scale engraved on its surface. The linear scale is usually 1 or 2 mm in length, subdivided into hundredths. The stage micrometer is used to directly measure the diameter of the field of view as well as to calibrate the ocular micrometer. Refer back to the diagrams in Chapter 1 (Fig. 1–2).

The *ocular micrometer* is a glass disk with a graduated linear scale etched on its surface. The graduations on the scale are usually from one to ten, and subdivided into tenths or fifths. The disk is placed at the focal plane of the ocular lens so that the image of the specimen and the scale of the ocular micrometer will both be in focus. (Refer back to Chapter 1 for the reasons why the ocular micrometer is so placed.)

The linear scale on the ocular micrometer cannot be given an absolute length value because individual differences in the magnification of the objective lens would cause the apparent length to change. It is, therefore, necessary to calibrate it against the stage micrometer. Why is it that the calibration data determined for one microscope will not be valid for any other microscope?

## APPLICATIONS

Once an ocular micrometer has been calibrated, its scale can be used to determine the length and diameter of a cell (or any other object that can be examined under the microscope). As we will see in later exercises, this information can then be used to calculate cell volume and cell weight. If the total number of cells in the suspension is known, the total biomass of those cells can be calculated.

If the area of the field of view is known (calculated from the diameter), the number of cells in a known volume of fluid can be determined. A portion of the sample, typically 0.01 ml, is carefully layered over a known surface area and allowed to dry. The cells, or any other particles, can be stained and then counted. Relatively simple mathematics are used to determine the number of cells distributed over the surface, based on the average number of cells in each field of view.

## OBJECTIVES OF THIS EXERCISE

The use of the microscope as a quantitative tool requires an understanding of how the measurements are made and why these calibrations are necessary. You should:

1. Understand the use of the ocular micrometer and how its relative units can be used to determine actual cell length and diameter.
2. Understand how the number of cells observed in the field of view can be related to number of cells on the surface being

examined or in the fluid spread over that surface.

## MATERIALS NEEDED FOR THIS EXERCISE

1. Microscopes equipped with ocular micrometers. The ocular micrometer should be mounted at the focal plane according to the specifications provided by the manufacturer of the microscope.
2. Stage micrometers. The stage micrometer should be carefully handled. Use only distilled water and lens paper to clean it; handle it only by its edges. Keep the micrometer in its case at all times (except, of course, when you are using it) and be careful that nothing is allowed to scratch its surface.

## LABORATORY PROCEDURES

### Part One—Determination of Diameter and Area of Field of View

The field of view can be measured by focusing the microscope on the stage micrometer and directly measuring its diameter. Study the examples shown in Fig. E21-1. Although not drawn to scale, the diagram clearly shows what you will observe when you focus each objective lens on the stage micrometer.

To improve accuracy, always use the same edge of the graduation mark as your reference line. As you increase the magnification, the width of the marks will appear to increase. You can avoid the error of including the thickness of the marks in your measurements by always using the same edge (right or left) of the marks.

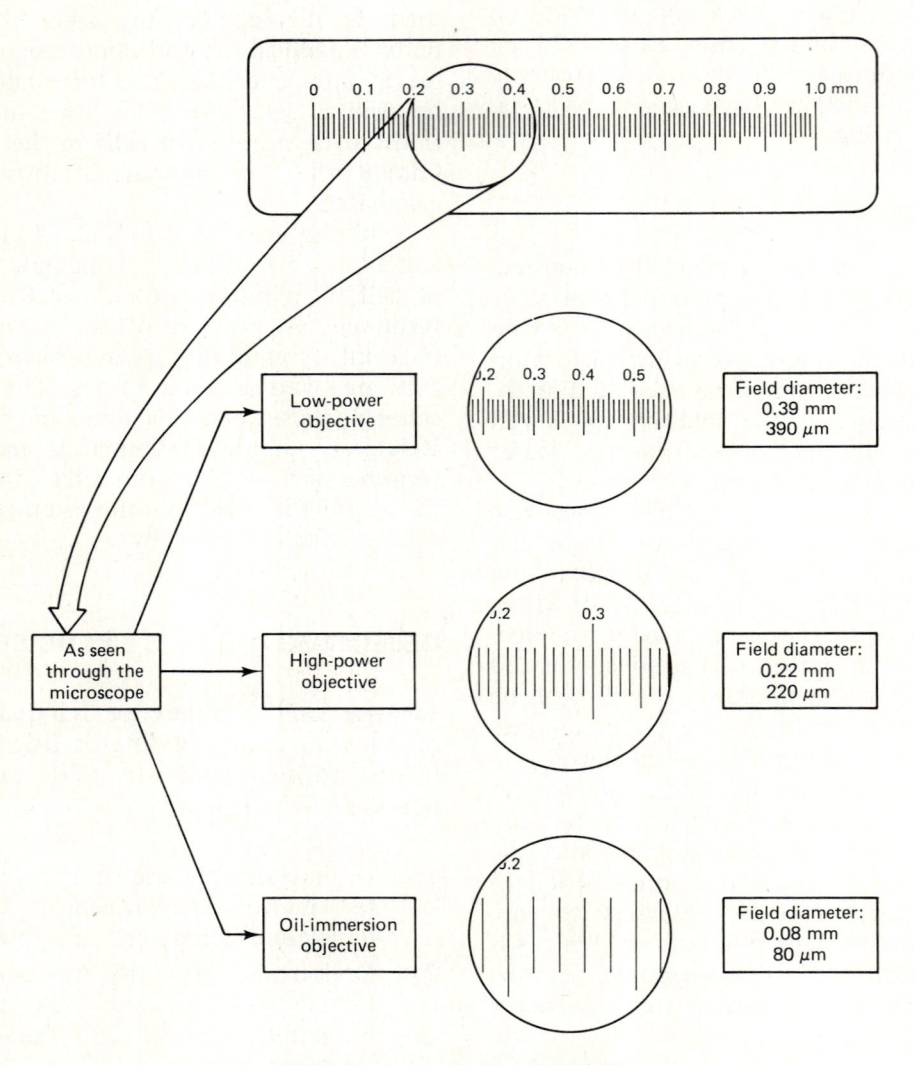

**Figure E21-1** Determination of field diameter.

1. Carefully align the stage micrometer on the microscope stage so that the graduated scale is approximately centered in the light path.
2. Center the low-power objective lens (10×) over the slide and carefully lower it until it is about ½ cm above the slide. If your microscope has the autostop feature, you should lower the lens until it stops.
3. Focus the image of the stage micrometer by moving the lens *away* from the stage. Alternately adjust the condenser and focus until a clear image is obtained.
4. Move the stage micrometer so that a convenient graduation is located against the left edge of the field of view. If your microscope does not have a flat field of view, you might need to adjust the focus so that the edges of the field of view are in sharp focus.
5. Determine the diameter of the field by observing the graduation closest to the right edge of the field. It might be necessary to extrapolate between graduations to determine the exact diameter.
6. Record the results in the appropriate blank on the Laboratory Report Form.
7. Repeat steps 3, 4, 5, and 6 for the other objective lenses on your microscope. Be particularly careful to ensure that there is no possibility of the lenses damaging the stage micrometer. Watch from the side when you change lenses. When you focus with the oil-immersion lens, use only the fine adjustment.
8. Calculate the area of the field of view for each magnification. Use the formula for the area of a circle (look it up). Record these results on the Laboratory Report Form.

**Part Two—Calibration of the Ocular Micrometer**

The ocular micrometer is calibrated by comparing its graduations with the graduations on the stage micrometer. Figure E21-2 shows, in idealized diagrams, how both micrometers might appear in a microscope field of view. In the example shown there are two points, indicated by arrows, where graduations on both micrometers coincide. When more than one such point is observed, use the longer distance for your calculations.

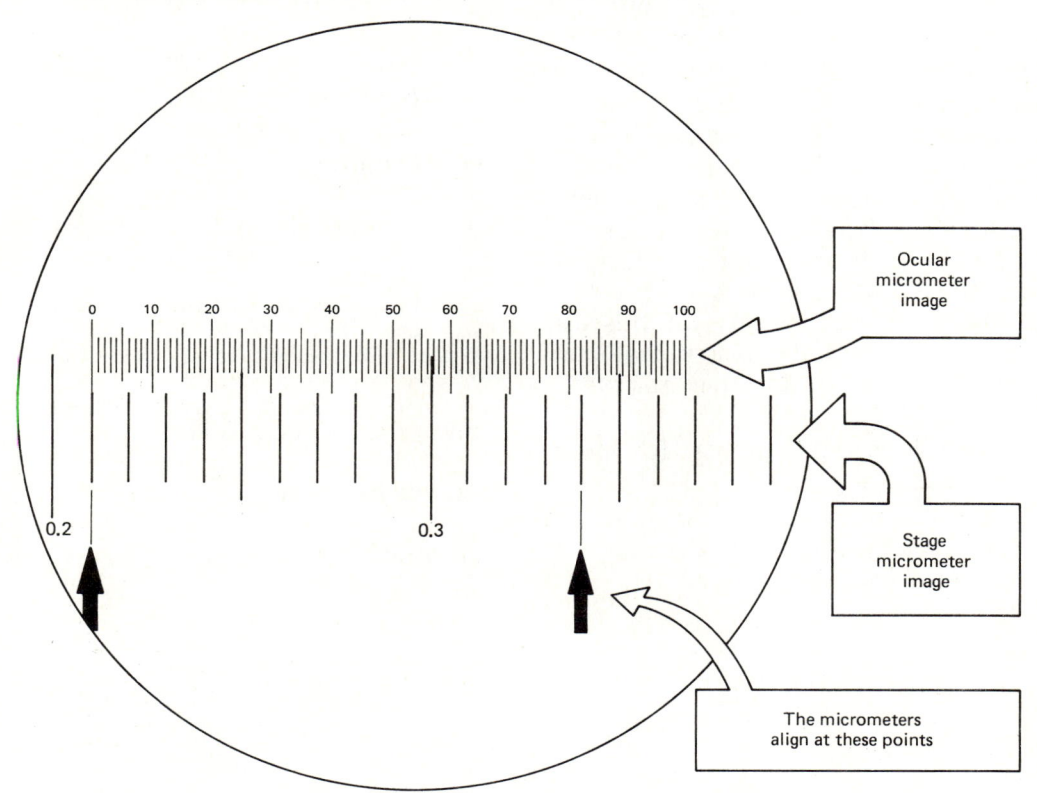

**Figure E21-2** Calibration of ocular micrometer.

1. Ascertain that the ocular micrometer is correctly installed in the ocular lens. The best way to do this is to focus on a specimen. If an ocular micrometer is properly installed, you will be able to see it in focus with the specimen. *There is no reason to remove the ocular lens.* If you do not see the ocular micrometer, ask your instructor for assistance.

    If you are using a binocular microscope you will need to determine which lens contains the ocular micrometer. As you look through the instrument, close one eye at a time. The lens with the ocular micrometer will become apparent.

2. Center the stage micrometer on the microscope stage.

3. As you look through the low-power objective lens, focus the microscope until the graduations of the stage micrometer appear in sharp focus. Rotate the ocular lens until the two sets of graduations are parallel to each other (see Fig. E21-2). It will be easier to complete the calibration if the two sets of graduations overlap somewhat.

4. Move the stage micrometer so that a convenient calibration mark is aligned with the zero mark on the ocular micrometer. Use the mechanical stage adjustments if you have them on your microscope.

5. Locate the point where another graduation on the ocular micrometer aligns with a graduation on the stage micrometer. In the example, there are two points of alignment, as indicated by the two arrows in the diagram. If more than one point of alignment is observed, use the one with the greatest length for your calculations. In this example, 82 units on the ocular micrometer align with 0.13 mm on the stage micrometer. (Note where zero is aligned.)

6. Record the alignment points for both micrometers in your notes.

7. Repeat steps 3, 4, 5, and 6 for each objective lens and record your data on the Laboratory Report Form.

8. Calculate the dimensions of the graduations on the ocular micrometer for each magnification. You can do this by dividing the length (in millimeters) observed on the stage micrometer by the number of units that coincide to that length on the ocular micrometer.

    Ocular micrometer unit length =

    $$\frac{\text{Stage micrometer length (in mm)}}{\text{Ocular micrometer units}}$$

    For example, in Fig. 21-2, you would divide 0.13 by 82.0, for a value of 0.0016. This indicates that each unit of the ocular micrometer is 0.0016 mm, or 1.6 $\mu$, in length.

    Ocular micrometer unit length =

    $$\frac{0.13 \text{ mm}}{82.0 \text{ units}} = .0016 \text{ mm}$$

9. Record all data in the Laboratory Report Form. This completes the calibration of your microscope. You will need to refer back to these data when you complete Exercises 22 and 23.

## REFERENCES

### Text References

1. Brock, Thomas D., David W. Smith, and Michael T. Madigan, *Biology of Microorganisms,* 4th ed. Chapter 7. Appendix 5.
2. Jensen, Marcus, *Introduction to Medical Microbiology.* Chapter 6.
3. Nester, Eugene W., et al., *Microbiology,* 3rd ed. Chapter 4.
4. Wistreich, George A., and Max D. Lechtman, *Microbiology,* 4th ed. Chapter 6.

### Resource References

1. Gerhardt, P., ed., *Manual of Methods for General Bacteriology.* Chapters 3 and 11.

# LABORATORY PROTOCOL

## Exercise 21—Calibration of the Microscope

Check each step when you complete it.

**Part One: Determination of Diameter and Area of Field of View**

_____  1. Obtain a stage micrometer and determine the length of the graduations etched on its surface. (It will be either 1 or 2 mm.)

_____  2. Center the stage micrometer on the microscope stage and carefully bring it into focus with the low-power objective lens.

_____  3. Using the mechanical stage (if your microscope is equipped with one) to position the stage micrometer, determine the diameter of the field of view at low power.

_____  4. Similarly, determine the diameter of the field of view for each of the other objective lenses. Record the data on the Laboratory Report Form.

_____  5. Calculate the area of each field of view and record those results in the appropriate blanks on the report form.

**Part Two: Calibration of the Ocular Micrometer**

_____  6. Determine that the ocular micrometer is correctly installed in the ocular lens of the microscope. If you are using a binocular microscope, determine which of the two ocular lenses the micrometer is installed in.

_____  7. Using low power, position the stage micrometer so that a convenient graduation mark aligns with the zero mark of the ocular micrometer. (Refer to Fig. 21-1.) Remember that the two scales may overlap somewhat to make it easier to determine the points of alignment.

_____  8. Locate another point where there is alignment of the two scales.

_____  9. Calculate the actual length of the units on the ocular micrometer.

_____ 10. Calibrate the ocular micrometer with the high-power objective.

_____ 11. Finally, focusing very carefully, calibrate the micrometer with the oil-immersion lens.

_____ 12. Record all data on the Laboratory Report Form.

# NOTES

Name: _____

## LABORATORY REPORT FORM

### Exercise 21—Calibration of the Microscope

1. Record the following information about your microscope:

    Serial number of microscope body: _____
    (or other identifying number)

    Serial number of low-power lens: _____

    Serial number of high-power lens: _____

    Serial number of oil-immersion lens: _____

2. Record the following data about the field of view of your microscope.

    | Lens | Diameter | | Area | |
    |---|---|---|---|---|
    | Low power: | _____ | (_____) | _____ | (_____) |
    | High power: | _____ | (_____) | _____ | (_____) |
    | Oil immersion: | _____ | (_____) | _____ | (_____) |
    
    Units in millimeters (micrometers)

3. Record calibration data for the ocular micrometer.

    | | Length of One Ocular Micrometer Unit | |
    |---|---|---|
    | Lens | Millimeters | Micrometers |
    | Low power: | _____ | _____ |
    | High power: | _____ | _____ |
    | Oil immersion: | _____ | _____ |

### ANSWER THE FOLLOWING QUESTIONS

1. Why must the ocular micrometer be located at the focal plane of the ocular lens?

2. Explain why you would not expect the actual diameter of the field of view observed at 1000 magnifications to be exactly one tenth of that measured at 100 magnifications.

3. Using your data, calculate how many fields of view would fit inside one square centimeter. Complete the calculations for each magnification.

        Low power (100X): _____
        High power (400X): _____
        Oil immersion (1000X): _____

4. If you determined that three spherical objects (like cells) had diameters equal to 1.65 ocular micrometer units at the three magnifications of your microscope, what would their actual diameters and area be?

| Lens | Diameter | Area |
|---|---|---|
| Low power: | | |
| High power: | | |
| Oil immersion: | | |

Units should be in millimeters or micrometers.

# Exercise 22
# Cell Dimensions: Measurement of Cell Size

The dimensions of a cell can be measured accurately if an ocular micrometer is installed in your microscope and if it has been correctly calibrated. In this exercise we will use the calibration data obtained in Exercise 21 to determine the dimensions of red blood cells and bacteria.

If the diameter and/or length of a cell is known, it is possible to calculate certain other biomass parameters, such as mean cell volume and mean cell surface area. These parameters have both clinical and ecological significance because if they can be accurately determined for individual cells, such data as total cell volume and total surface area for the population can be calculated. Also, the weight of a cell can be estimated if the volume and density are known.

## SOME BACKGROUND AND ASSUMPTIONS

Most cells are spherical in shape and can be assumed, for the purposes of this exercise, to be round, spherical bodies, not unlike tennis balls or basketballs. Rod-shaped cells can be assumed to resemble cylinders whose length and diameter can be measured. Finally, red blood cells, because of their concavity, are unique. Except for their diameter, the biomass parameters of red blood cells cannot be calculated easily.

The formulas needed to calculate the biomass parameters for cocci and rod-shaped cells are relatively simple. If cocci are treated as spheres, then the formulas used for circular and spherical bodies can be used. If rods are being examined, the length of the rod and its circumference (calculated from diameter) can be used to determine surface area and volume.

Before you begin this exercise, you should look up the formulas for, or figure out how to calculate, the following:

1. Area of a circle:

2. Circumference of a circle:

3. Volume of a sphere:

4. Surface area of a sphere:

5. Surface area of a cylinder:

6. Volume of a cylinder:

*Hint:* When you calculate the weight and volume of cells, you must consider the conversion of specific gravity from grams per milliliter to micrograms per microliter, and volume relationships from cubic microns to microliters. They are not a one-to-one conversion. Can you figure it out?

## OBJECTIVES OF THIS EXERCISE

Much information about microbial populations, and about individual cells in those populations, can be derived from simple diameter and length measurements. In this exercise you should:

1. Become familiar with procedures for measuring the length and diameter of cells.

2. Understand the relationship between cell diameter and length and cell volume and surface area.

3. Understand the relationship between cell volume and cell weight.

## MATERIALS NEEDED FOR THIS EXERCISE

1. Microscopes equipped with ocular micrometers. If possible, use the microscope you calibrated in Exercise 21. If that is not possible, either obtain the data for the microscope you are using or calibrate the one you will use for this exercise.
2. Twenty-four-hour broth cultures of:
   *Bacillus cereus*
   *Staphylococcus aureus*
3. Small test tubes containing about 5.0 ml of 0.9% saline.
4. Lancets.
5. Alcohol wipes.
6. Pasteur pipettes.
7. Glass slides and coverslips.

## LABORATORY PROCEDURES

### Part One—Diameter of Red Blood Cells

1. Prepare a suspension of red blood cells. Carefully cleanse a fingertip with an alcohol wipe. After the skin air dries, quickly pierce the skin with a sterile lancet. Allow several drops of blood to fall into the saline in one of the small tubes. Hold a piece of cotton or gauze against the puncture until blood clotting is complete.
2. Make a wet mount (Exercise 2) of the red blood cell suspension. Allow a few minutes for the cells to settle, then examine them under the low power of the microscope.
3. Manipulate the ocular micrometer and the slide until you can measure the diameter of a red blood cell. (See Fig. E22-1.) Align one edge of the cell against a convenient graduation and determine the cell's diameter in ocular micrometer units. You should repeat this several times to ensure that the cells are approximately the same size. Convert these figures to actual diameters, in millimeters or micrometers.
4. Repeat step 3 with the next higher magnification to determine if a more accurate measurement can be obtained.
5. Calculate the mean cell diameter of the red blood cells.

Red blood cells          Bacteria

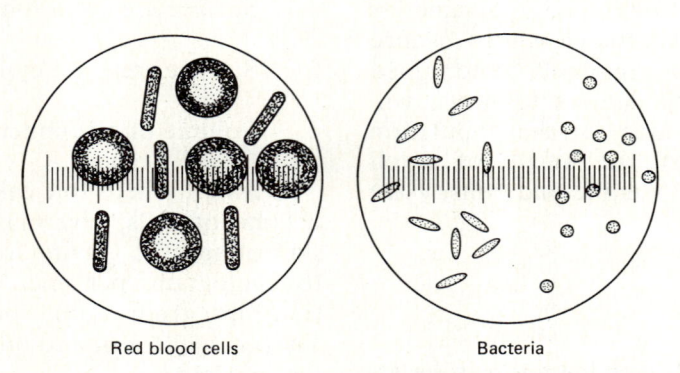

**Figure E22-1**   Measurement of cell diameter (idealized drawing).

6. Attempt to measure the thickness of red blood cells.
7. Record all your results in the Laboratory Report Form.

### Part Two—Diameter and Length of Bacteria

Bacterial cells are easier to see, and therefore easier to measure, if they are stained. However, the drying associated with the preparation of a smear almost always causes some distortion, usually shrinkage, of the cells. In this exercise, however, we will assume that there is only negligible distortion introduced by smear preparation and the simple staining procedures. As an alternate procedure, consider using wet mounts to measure the length and diameter of the cells.

Although rod-shaped cells can be viewed as cylinders, with the diameter of the cell usually remaining constant (except for diphtheroids and ellipsoidal cells) over its length, we must decide how we will account for the ends of the cell. The simplest procedure is to assume the ends are not significantly rounded and can, with only a small amount of error, be considered flat. We will consider the length of a rod-shaped cell to be the total length, including any rounding of the ends. (See Fig. E22-2.) This problem is not a serious one if we are only concerned with the length of the cell, but when we want to calculate biomass, we must consider how much error this assumption is likely to introduce.

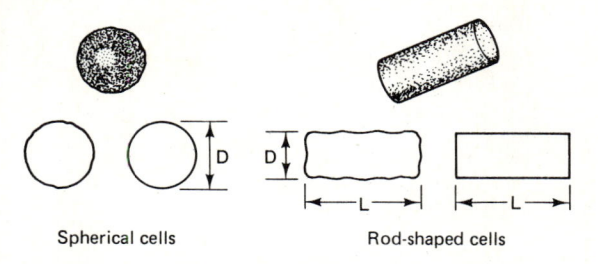

**Figure E22-2** Dimensions of spherical and rod-shaped cells.

1. Prepare simple stains (Exercise 2) of the two bacterial suspensions. It is particularly important that you prepare smears with well-separated cells. If necessary, dilute the bacterial suspension with saline.
2. As an alternate procedure, add one loopful of crystal violet to 1.0 ml of a 24-hour broth culture, and use this preparation to make wet mounts.
3. Using the oil-immersion objective, determine the mean diameter of *Staphylococcus aureus* cells. You should take at least 10 measurements to account for any variation in the diameter of the cells.
4. Similarly measure the length and diameter of *Bacillus cereus*. Again, take at least 10 measurements and calculate the mean. It is good practice to measure the length and diameter of the same cell and to keep the two measurements paired. (Why?)
5. Record all your results in the Laboratory Report Form.

## Part Three—Calculation of Cell Volume, Surface Area, and Weight

1. Calculate the surface area of the average *Staphylococcus aureus* cell. Use the formula for the surface area of a sphere and assume $\pi$ to be 3.14.
2. Calculate the volume of your average *Staphylococcus aureus* cell. Then, assuming that the density of bacterial cytoplasm is equal to water (1.0), calculate the weight of the average cell in the suspension.
3. The calculation of the surface area of the average *Bacillus cereus* cell requires that you multiply the circumference of the cell by its length and then add the area of two ends. Use the diameter of the cell to calculate the area and circumference of a circle.
   a. Calculate the area of the walls of the cylinder (cell): Multiply the circumference by the length of the cell.
   b. Calculate the area of the ends of the cylinder (cell): Calculate the area of the circle and multiply by two.
   c. Add the results of a. and b. together to obtain the surface area of the average *Bacillus cereus* cell in the suspension.
4. Calculate the volume of the rod-shaped cell by multiplying the area of the ends of the cell by the length of the cell. The weight of the cell can now be calculated.

## REFERENCES

### Text References

1. Brock, Thomas D., David W. Smith, and Michael T. Madigan, *Biology of Microorganisms,* 4th ed. Chapter 7.
2. Jensen, Marcus, *Introduction to Medical Microbiology*. Chapter 6.
3. Nester, Eugene W., et al., *Microbiology,* 3rd ed. Chapter 4.
4. Wistreich, George A., and Max D. Lechtman, *Microbiology,* 4th ed. Chapter 6.

### Resource References

1. Gerhardt, P., ed., *Manual of Methods for General Bacteriology*. Chapters 3 and 11.

# LABORATORY PROTOCOL

### Exercise 22 — Cell Dimensions: Measurement of Cell Size

Check each step when you complete it.

_____  1. Determine that your microscope is equipped with an ocular micrometer. Obtain the calibration data for that microscope. It is preferable to use the same microscope that you calibrated in the previous exercise. Record the actual length of one ocular micrometer unit.

| LENGTH OF ONE OCULAR MICROMETER UNIT | |
|---|---|
| With low power: | _____ |
| With high power: | _____ |
| With oil immersion: | _____ |

### Part One — Diameter of Red Blood Cells

_____  2. Prepare a suspension of red blood cells in 0.9% saline. Use proper cleansing procedures for the tip of your finger.

_____  3. Measure the diameter of at least ten cells and calculate the mean diameter. Convert these from ocular micrometer units to actual length in millimeters and micrometers.

_____  4. Attempt to measure the thickness of at least ten cells and similarly determine the mean actual thickness in millimeters and micrometers.

_____  5. Record these data in the Laboratory Report Form.

### Part Two — Diameter and Length of Bacteria

_____  6. Make smears and simple stains of the bacterial cultures of *Staphylococcus aureus* and *Bacillus cereus*. Be careful to ensure that the cells are well isolated.

_____  7. As an alternate procedure, transfer one loopful of crystal violet to 1 ml of 24-hour broth culture and make a wet mount (Exercise 2) of the stained suspension.

_____  8. Measure the diameter of the cocci. Take sufficient measurements so that the mean diameter can be calculated.

_____  9. Similarly, determine the mean diameter and mean length of the *Bacillus cereus* cells in the culture. Be sure to take the diameter and length measurements on the same cells.

_____  10. Record these data in the Laboratory Report Form.

### Part Three — Cell Volume, Surface Area, and Weight

_____  11. Calculate the surface area and volume of *Staphylococcus aureus*.

_____  12. Calculate the surface area and volume of *Bacillus cereus*.

_____  13. Calculate the mean cell weight for each type of bacterium. Assume that bacterial cytoplasm has a density of 1.0 gm/cc.

_____  14. Record all data in the Laboratory Report Form.

**NOTES**

Name: _____

## LABORATORY REPORT FORM

**Exercise 22 — Cell Dimensions: Measurement of Cell Size**

1. Complete the following table.

**CELL MEASUREMENTS**

| | |
|---|---|
| Red blood cell diameter: | _____ |
| Thickness of red blood cells: | _____ |
| Diameter of *Staphylococcus aureus*: | _____ |
|   Wet mount measurements: | _____ |
| Diameter of *Bacillus cereus*: | _____ |
|   Wet mount measurements: | _____ |
| Length of *Bacillus cereus*: | _____ |
|   Wet mount measurements: | _____ |

**CALCULATED VALUES FOR BACTERIAL CELLS**

*Staphylococcus aureus*

| | Dry | Wet |
|---|---|---|
| Surface area of average cell: | _____ | _____ |
| Volume of average cell: | _____ | _____ |
| Weight of average cell: | _____ | _____ |

*Bacillus cereus*

| | Dry | Wet |
|---|---|---|
| Surface area of average cell: | _____ | _____ |
| Volume of average cell: | _____ | _____ |
| Weight of average cell: | _____ | _____ |

## ANSWER THE FOLLOWING QUESTIONS

1. What would be the total cell surface area in a 10-ml suspension of *Bacillus cereus* that contained 100,000 cells per milliliter?

2. What volume would those cells occupy and what would they weigh?

3. How would you determine the volume of a red blood cell? (*Hint:* You need to know the total volume of a large number of cells.)

4. Did you experience any difficulty measuring the thickness of the red blood cells? If you did, what was the reason for the difficulty?

# Exercise 23
# Bacterial Population Counts: Microscope Counting Methods

In the last exercise we learned that the microscope may be used to measure the size of cells, and that once the length and/or diameter was determined, the volume, weight, and surface area of the cells could be calculated. In this exercise we will learn how to use the microscope to count the number of cells in a population.

There are two general methods of determining the size of cellular populations. One method relies on the actual counting of the cells using specially prepared microscope slides, counting chambers, or electronic particle counters. In the second method, the cells in the population are allowed to grow and we count the resulting colonies or measure some other biological activity (such as increased protein, carbon dioxide production, or acid production). In Exercise 24 you will determine the size of a bacterial population by the plate count method, but in this exercise, you will use direct microscope counting methods to actually count red blood cells and bacterial cells.

## DIRECT MICROSCOPE COUNTING

When more than one method is available to determine what appears to be the same data (e.g., the size of the population), you must often decide which of the methods is best. The answer to this question frequently lies in an understanding of the limitations of the methods and an understanding of what each of those methods actually measures. There certainly are many circumstances where plate counting will give all the information you need, but just as certainly, there are circumstances where microscope counting methods provide the quickest and best route to your data. So, to answer the question raised in the first sentence, you must choose the method that gives you the most useful data.

Microscope counting methods often provide information that is just not possible to obtain in any other way. Some examples:

- A microscope count includes *all* cells in the population, whereas a plate count includes only those cells that are capable of growing under the environmental conditions that were used in the experiment—medium, temperature, pH, oxygen content, and so on. For example, anaerobic bacteria will be unable to grow using any of the methods that incubate the cultures in air. Similarly, if you were to use anaerobic methods, you would effectively preclude growth of the aerobic bacteria.

- A plate count cannot, of course, determine the number of nonviable cells in the suspension. For example, if you complete plate counts on a milk sample before and after pasteurization, you should get very different results. A comparison of the counts obtained by microscope and plate-count methodology would yield some information about the quality of the milk before pasteurization and about the effectiveness of the pasteurization process.

- A microscope count will sometimes be the only practical way of determining the size of the population or of measuring the number of cells in a given volume. A blood count is a good example.

## DIRECT MICROSCOPE COUNTING PROCEDURES

### Counting Chambers

A counting chamber is a specially constructed glass slide that has a counting grid of known dimensions etched on its surface. A coverslip is supported a known distance above the surface that the counting grid is etched on. Examine Fig. E23-1. Notice the location of the counting chamber and the counting grid.

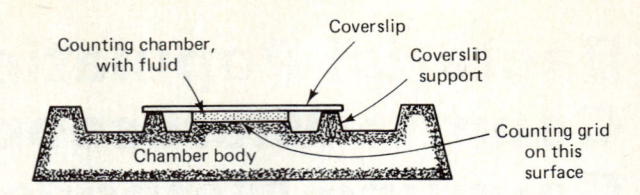

**Figure E23-1** Counting chamber, cross-section.

Figure E23-2 shows, in a composite diagram, a counting-chamber grid, with Neubauer ruling, and how it might look if you were counting red blood cells or bacteria. The square in the

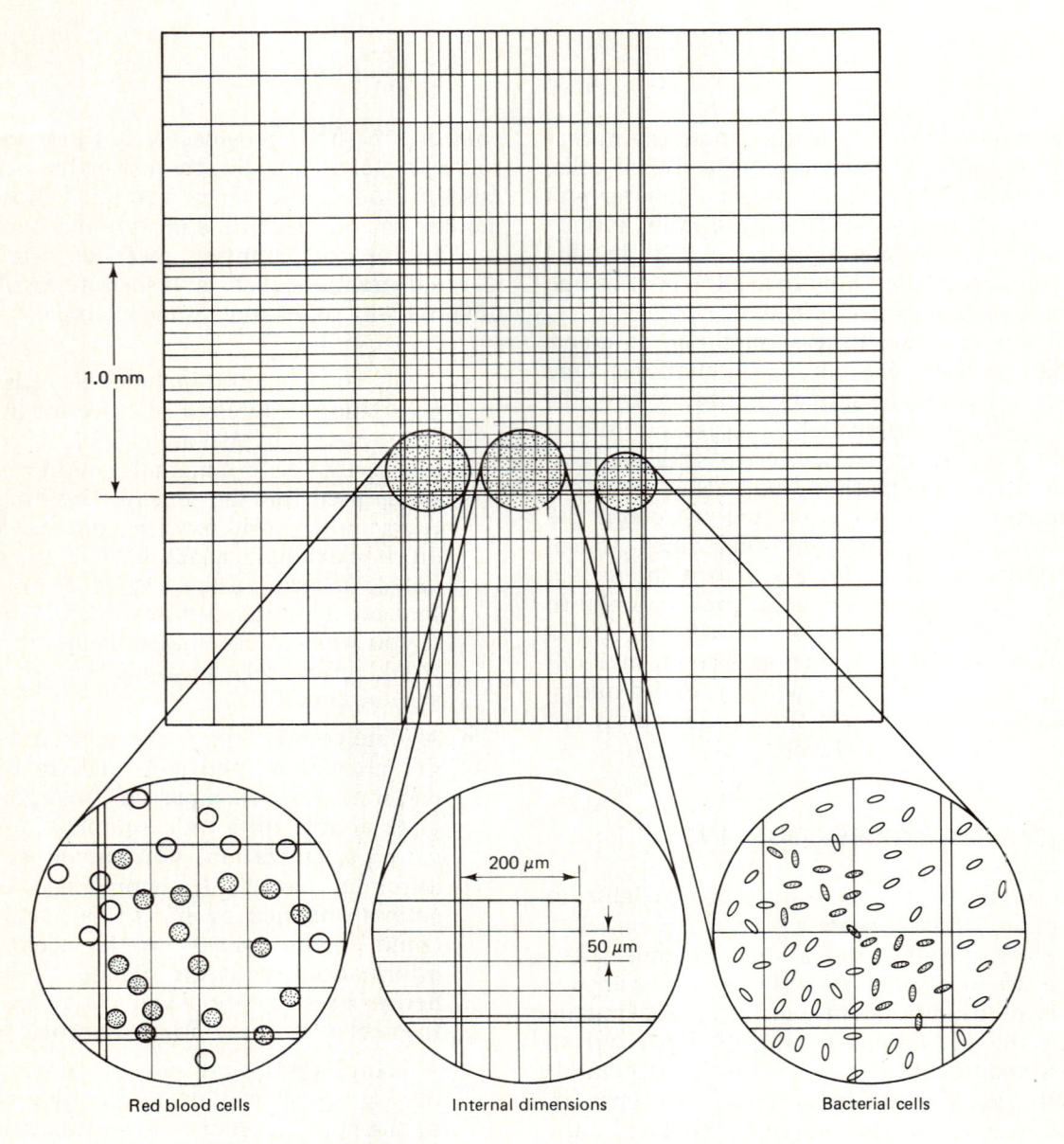

**Figure E23-2** Cell-counting chamber with exploded views.

center of the grid, bounded by the double lines, is 1 mm². The volume of the fluid in the area bounded by the grid can be calculated because the dimensions of the grid and the space above it are known. Think of the counting area as a chamber bounded on the top by the coverslip, on the bottom by the chamber itself, and on the sides by the dimensions of the counting grid.

The center grid is subdivided into smaller units, which, depending upon the number and size of the cells in the fluid being counted, may also be used for counting. For example, if you were counting bacteria, you would use the smaller squares (bounded by the thin single lines), but if you were counting red blood cells, the larger squares (bounded by the double and/or heavy lines) would be used. And, as you can see in the diagrams, different magnifications may also be used.

The center square, bound by the double lines, has an area of exactly 1 mm². This center square is subdivided into 25 smaller squares, each of which has sides 0.20 mm (or 200 µm) long. They are bounded by heavy lines and are 1/25 mm². The smallest squares (bound by thin lines) have sides that are 0.05 mm (50 µm) long and which are 1/400 mm².

In practice, you would count the cells in several squares (at least five), calculate the average number of cells in each square, and then multiply by the appropriate numbers to obtain the number of cells per milliliter. You need to know the volume of the fluid being counted in the chamber (multiply the area of the grid by the depth of fluid over the grid) and the dilution of the fluid used to fill the counting chamber (red blood cells are usually counted at a dilution of 1:200).

Typically, the counting-chamber grid used to count bacteria or red blood cells is the central, 1-mm² grid. The depth of the space above the grid in a Neubauer chamber is 0.10 mm. The volume of a chamber, with a depth of 0.10 mm, is 0.10 mm³.

In making the actual count, you should ignore the cells touching the upper and left borders and count all the cells that touch the lower and right margins (even if most of the cell appears to lie on the wrong side of the line).

In the example shown in Fig. 23-2, there are 17 countable (shaded) red blood cells in the section of the counting grid shown (five are touching the upper and/or left margin). The countable bacterial cells in two sections are also shown (11 cells in each square).

### Direct Microscope Count

In the direct microscope count, a known volume of the suspension is smeared on a slide over a known surface area, usually 1 cm². After the suspension dries, the cells can be stained and counted. This method is commonly used to count bacteria in milk and is often referred to as the *Breed count*. In this procedure, 0.01 ml of milk is spread over an area of exactly one cm². The number of bacteria in several randomly chosen fields of view are counted and the mean number of bacteria per field calculated. If you know the area of the field of view (from Exercise 21), you can calculate the number of bacteria in the square centimeter over which the 0.01 ml of suspension was spread. This is, of course, the number of bacteria in 0.01 ml of the suspension you are counting. Multiply by 100 to obtain the number of bacteria per milliliter.

## OBJECTIVES OF THIS EXERCISE

The microscope can be used to count the number of cells in a suspension. To do this, you must:

1.  Understand the use of counting chambers and the relationship between the surface area of a grid and the volume of the space above it.
2.  Understand how the number of cells in a volume of fluid can be determined by the Breed-counting procedure.
3.  Understand that the direct-counting method will often, but not always, provide data that cannot be obtained by the plate-count method.
4.  Appreciate that most experimental methods often have limitations that preclude their use in all instances.

## MATERIALS NEEDED FOR THIS EXERCISE

1.  Cell-counting chambers, with Neubauer or Petroff-Hauser ruling. Regardless of the type of chamber used, be sure to determine

the dimensions of the counting grid (usually 1.0 mm²) and the subdivisions contained therein. Also determine the depth of the counting chamber (either 0.10 mm [Neubauer] or 0.02 mm [Petroff-Hauser]).
2. Breed-counting slides. If breed-counting slides are not available, you can mark off 1 cm² on a glass slide with a wax marking pencil.
3. Pipettes capable of accurately delivering 0.01 ml.
4. Red blood cell dilution pipettes. If these are not available, you will need pipettes and diluent to make a 1:200 dilution of whole blood. As a last resort, you can assume that one drop in 10.0 ml will give about a 1:200 dilution.
5. A 24-hour broth culture of bacteria. Almost any *nonmotile* bacterium will suffice. Alternately, you might use some milk that is unpasteurized or which has been kept for a few hours at room temperature.
6. Lancets and alcohol wipes.
7. Test tubes containing 9.0 ml of 0.9% saline.

## LABORATORY PROCEDURES

### Part One: Counting Chambers

1. Obtain a counting chamber and the coverslip that is to be used with it. (Why is this coverslip so much thicker than other coverslips you might have used?) Carefully clean the counting surface and the coverslip. Use only distilled water for a solvent.
2. Using the lancet and alcohol wipes, puncture the tip of your finger and collect sufficient blood to complete a red blood count.
   If you are using a red blood cell diluting pipette or other blood-diluting device, follow the instructions for that device. If you are making your own diluting system, transfer 0.01 ml of blood into 1.99 ml of 0.9% saline.
3. If Petroff-Hauser counting chambers are available, mix one loopful of crystal violet with 1.0 ml of the 24-hour broth culture and use this in place of the red blood cell suspension.
4. Draw a small amount of the diluted red blood cell or stained bacterial cell suspension into a pipette. Allow some of the suspension to form a bead on the pipette tip. Touch the tip (with the small bead of fluid) to the edge of the counting chamber. (Some counting chambers have a triangular groove cut into their surface for the pipette tip.)
5. The droplet will be drawn into the chamber by capillary action and should just fill it. If there is too much fluid in the chamber, it will spill over into the grooves. Any excess fluid should be removed with absorbent paper.
6. Allow the counting chamber to stand for a few moments.
7. Gently place the counting chamber on the microscope stage, with the counting chamber centered over the light source. Locate the grid with the low power of the microscope and position it so that you can count the number of cells in one of the squares. Change magnification as needed.
8. Count the number of cells in at least ten squares and calculate the mean number of cells per square.
9. Complete the necessary multiplications to determine the number of cells per cubic millimeter and cubic centimeter of blood.
   a. Multiply the number of cells per square (step 8) by the number of squares in the grid. This will give you the number of cells in the volume of the counting chamber (either 0.10 ml [Neubauer] or 0.02 ml [Petroff-Hauser]).
   b. Multiply that answer by either 10 (Neubauer chamber) or 50 (Petroff-Hauser chamber) to obtain the number of cells per milliliter in the *diluted* sample.
   c. Finally, multiply by the dilution (usually 200 for red blood cells) to obtain the number of cells per milliliter.
10. Record your calculations and results in the Laboratory Report Form.

### Part Two: Breed-Counting Procedure

1. Obtain a Breed-counting slide. It will contain from one to five marked staining areas. Each area is 1 cm². If a Breed slide is not available, you can approximate one by marking off 1-cm squares with a wax pen-

cil. Use a centimeter ruler to draw a square centimeter on a piece of paper. Gently warm a clean slide over the Bunsen burner and then position it over the square. Quickly (before the slide cools) outline the square centimeter on the slide with a wax pencil.

You may avoid using the wax pencil by using a loop to spread the drop of fluid over a 1-cm$^2$ area (use the drawing on the paper as a template).

2. Use the 0.01-ml pipette to transfer 0.01 ml of the bacterial suspension (or milk) to one of the squares on the slide. If the slide is clean, the drop will spread evenly over the glass and fill the marked-off area. If necessary, use a loop to spread the fluid.
3. Make both a 1:10 and a 1:100 dilution of the bacterial suspension and repeat step 2, applying the suspensions to different staining squares.
4. Allow the smears to air dry. Heat fix the smears over a boiling-water bath for a few minutes. If you are using milk for the count, flood the slide with xylol and then rinse gently with alcohol (95%). The xylol removes the fat in milk and the alcohol washes any residual xylol from the slide.
5. Stain the smears with methylene blue for about 30 seconds.
6. Decolorize gently with alcohol until the smears appear light blue. Rinse briefly with distilled water.
7. Allow the slides to air dry before attempting to count the bacteria.
8. Using the oil-immersion objective, count the number of bacteria in each of at least ten fields of view. Calculate the mean number of bacteria per field.
9. Calculate the number of bacteria per milliliter of milk:
   a. Determine the number of fields of view in 1 cm$^2$. Divide 1 cm$^2$ by the area of the field of view (Exercise 21).
   b. Multiply the number of fields/cm$^2$ by the number of bacteria/field to obtain the number of bacteria per 0.01 ml. Then multiply by 100 to calculate the number of bacteria per milliliter of bacterial suspension (or milk).
10. Record your calculations and results in the Laboratory Report Form.

## REFERENCES

### Text References

1. Brock, Thomas D., David W. Smith, and Michael T. Madigan, *Biology of Microorganisms*, 4th ed. Chapter 7.
2. Jensen, Marcus, *Introduction to Medical Microbiology*. Chapter 6.
3. Nester, Eugene W., et al., *Microbiology*, 3rd ed. Chapter 4.
4. Wistreich, George A., and Max D. Lechtman, *Microbiology*, 4th ed. Chapter 6.

### Resource References

1. Gerhardt, P., ed., *Manual of Methods for General Bacteriology*. Chapters 3 and 11.

# LABORATORY PROTOCOL

## Exercise 23 — Bacterial Population Counts: Microscope Counting Methods

Check each step when you complete it.

### Part One: Counting Chambers

_____ 1. Obtain a counting chamber and determine the dimensions of the central grid and the depth of the chamber (Neubauer chamber or Petroff-Hauser chamber).

_____ 2. If you are using a Neubauer chamber, you should count red blood cells. Using proper techniques, take a sample of blood and prepare a 1:200 dilution of the blood. If available, use a red blood cell counting pipette or other diluting device.
   If a Petroff-Hauser chamber is available, prepare a stained suspension of bacteria (one loopful of crystal violet in 1.0 ml of a 24-hour broth culture).

_____ 3. Carefully fill the counting chamber. (See steps 3, 4, and 5, page 224.)

_____ 4. Position the counting chamber on the microscope slide and locate the part of the grid you wish to use for counting the cells.

_____ 5. Count the number of cells in at least five squares and calculate the mean number of cells per square.

_____ 6. Complete the necessary calculations to determine the number of cells per milliliter of whole blood. Record your calculations and results in the Laboratory Report Form.

### Part Two: Breed-Counting Procedure

_____ 1. Obtain a Breed-counting slide or prepare a facsimile slide as explained in the exercise.

_____ 2. Place exactly 0.01 ml of the bacterial suspension or of milk in one of the squares. If necessary, use the inoculating loop to spread the fluid over the surface of the counting square.

_____ 3. Heat fix and stain the slides according to the directions given in this exercise.

_____ 4. Examine the slide with the oil-immersion objective and count the bacteria in at least ten fields of view. Calculate the mean number of bacteria per field of view.

_____ 5. Complete the necessary calculations to determine the number of bacteria in each milliliter of the bacterial suspension or milk. Remember that you will need to look up the area of the field of view that you calculated in Exercise 21.

_____ 6. Record all calculations and results in the Laboratory Report Form.

# NOTES

Name: _____

## LABORATORY REPORT FORM

**Exercise 23 – Bacterial Population Counts: Microscope Counting Methods**

1. Record the following information about the counting chamber you used.

    Type of counting chamber: _____
    (Neubauer or Petroff-Hauser)
    Length of one side of the center grid: _____
    (usually bound by double lines)
    Area of center grid: _____
    Depth of counting chamber: _____
    Volume of space over center grid: _____
    Length of side of smallest squares: _____
    (usually bound by thin lines)
    Number of these per $mm^2$: _____
    Length of side of medium squares: _____
    (usually bound by heavy lines)
    Number of these per $mm^2$: _____

2. How many red blood cells were there in each milliliter of blood? Show all calculations. Skip this question if you counted a bacterial suspension instead of red blood cells.

3. How many bacterial cells were there in each milliliter of the bacterial sample you counted? Show all your calculations. If you used a Petroff-Hauser chamber to count a bacterial suspension, show those calculations also.

## ANSWER THE FOLLOWING QUESTIONS

1. List three circumstances where a direct microscope count would be preferred over a plate count.

2. Why is lack of sensitivity one of the most critical disadvantages of the direct microscope count?

3. If you erroneously counted your cells so that you inadvertently included one extra cell, how much error (in terms of extra cells) would your results include? Use your own data for both red blood cells and bacteria.

4. Assuming your bacteria were cocci, with a diameter of 1 µm, what would be the total volume, total surface area, and total weight of all of the bacteria in your sample? (Don't forget to estimate the volume of the sample!) Show all calculations. (*Hint:* If you know the mean values for these measurements, you can calculate the totals if you know the number of cells in the population.)

   Assumptions:

   $$\text{Specific gravity} = 1 \text{ gm/ml } (1000 \text{ µg/µl})$$

   $$\text{One µl} = 10^9 \text{ (i.e., } 1 \times 10^9\text{) µm}^3.$$

# Exercise 24
# Bacterial Population Counts: Viable Cell Counts

The size of a bacterial population can be estimated by measuring some biological activity of the population or by plating the cells on a suitable medium and allowing them to grow into visible colonies. The term *viable cell count* refers to the counting of cells by plating them on a nutrient medium and counting the colonies that develop. Often the results are referred to as *plate counts* or *colony counts* and the results reported as *colony-forming units*.

Colony counts have some distinct advantages when compared to direct microscope counts. The colony count is significantly more sensitive than the direct count. Some microbiologists claim that there must be more than 50,000 bacteria in each milliliter before you can reliably and accurately count them on a microscope slide. On the other hand, plate-counting procedures, particularly when membrane-filtration techniques are used, can detect a very few colony-forming units in relatively large volumes of liquid. Of course, if hand counting is used, you must have at least 30 colonies on the plate before the count is considered to represent a valid sample.

## CONSIDERATIONS USED IN PLATE COUNTING

Certain assumptions and considerations have to be taken into account in all plate-counting procedures. These include:

1. Each colony is assumed to be the progeny of a single cell. If we could be assured that each colony did arise from a single cell, then the assumption that the number of colonies equals the number of cells in the suspension could be accepted. In fact, while it is true that some types of bacteria do readily dissociate into single cells, many do not.

Many microbiologists prefer to use the term *colony-forming units per milliliter* (CFU/ml) instead of *bacteria per milliliter*. They point out that it is the ability of the cells to grow and to exercise their usual biological activity that is of concern. It really doesn't matter whether the colony (or infectious unit) arose from a single cell or not; what does matter is that it will grow into a colony or may cause an infection.

The variability in the tendency of bacterial cells to remain clumped will introduce some error to your counts unless you are careful to treat each sample the same. For example, you should mix the dilution blanks in a consistent manner (e.g., 50 shakes, one minute on a vortex mixer at a given speed, etc.).

2. Except for relatively rare and unusual circumstances, plate counts of naturally occurring populations actually select for relatively small segments of these populations. The growth medium that is used in the petri plates as well as the conditions under which the plates are incubated are themselves selective.

Pure cultures of laboratory strains of bacteria, under some circumstances, represent an example where all cells might be expected to grow when transferred to a petri plate containing an appropriate nutrient medium.

Most natural populations of bacteria, such as would be found in the intestinal tract, in water samples, or in soil samples, contain bacteria that are not able to grow in the environmental conditions used for plate counting.

Often, special media must be used, or other environmental conditions must be manipulated. These include light, pH, anaerobiosis, temperature, special sources of oxidizable carbohydrate, and so on.

3. Frequently, however, the selectivity of the medium or the growth environment can be used to encourage the growth of specific kinds of bacteria. For example, we might want to know how many anaerobic bacteria were in the population. In soil microbiology it is important to know how large the cellulose-decomposing segment of the population is. In water, we need to know the number of coliforms before we can determine whether the water is safe to drink.

In clinical microbiology, the presence of more than 100,000 bacteria in urine is considered significant (Exercise 12), but in most other cases, where a normal flora is usually present, we are only interested in certain types of bacteria (beta-hemolytic, lactose nonfermenters, etc.).

4. The application of membrane-filtration techniques to plate-counting procedures makes it possible to greatly increase the sensitivity of such counts. For example, it is possible to pass large volumes of fluid through a filter, trapping the bacteria on the filter. If the filter is then placed on a filter-paper pad that has been saturated with medium and then incubated, the bacterial cells will grow into distinct colonies. In theory, any bacteria (even one) in the entire volume of fluid that passed through the filter will appear as a colony after suitable incubation. The technique is not only very sensitive, but the results can be obtained sooner. These techniques have found wide application in the quality testing of water, foods, and medicines.

5. Finally, it is often necessary to perform both a direct microscope count *and* a plate count on a sample. What would you conclude from a milk sample that had a plate count of 1,000 bacteria/ml and a direct count of 25,000 bacteria/ml? What would a growth curve look like if you plotted both plate counts and direct counts? (Why not plan an experiment to find out?)

## SOME REMINDERS

Plate counts usually require that the sample be diluted before plating. Study the diagram in Fig. 24-1, and, if necessary, review the material in Chapter IV. The reason for the dilution of the sample is to ensure that you will have plates with between 30 and 300 colonies on them. As a starting point, a 24-hour culture of *E. coli* will contain a cell population of between 1 and 100 *million* cells per milliliter.

## SOME APPLICATIONS

While it is often important to know the size of a bacterial population, it is sometimes even more important (and interesting) to learn how that population changes. For example, what would be the effect of changing the nutrients in a medium? How would you go about testing which medium produced better growth—nutrient broth or brain-heart-infusion broth?

What results would you expect to obtain if you did a plate count on a broth culture every two or three hours for about 24 hours? Would you be able to demonstrate the typical growth curve? Why not try it?

How would you evaluate the quality of water or milk? Most foods must meet standards that include limits on the numbers of bacteria per gram or milliliter. Often they will specify limits that include total bacteria as well as total coliforms. How would you modify this exercise to detect coliforms? (*Hint:* Refer to Exercise 13.)

## OBJECTIVES OF THIS EXERCISE

Viable cell counts determine the number of living cells in a population. Although this information may be most important, there are some limitations on the procedure. In this exercise you should:

1. Understand the relationship between the serial dilution and the plating of the sample.
2. Be able to list several of the limitations of the plate-count procedure.
3. Understand the limitations and advantages of membrane-filter counts.
4. Be able to discuss how plate counting and direct counting can often provide different, yet complementary, information about the population being studied.

## MATERIALS NEEDED FOR THIS EXERCISE

### Part One: Pour-Plate Counts

1. Sufficient dilution blanks to dilute the sample to 1:10 million. Seven tubes containing 9.0 ml of water or saline (0.9%) will be required if a simple tenfold serial dilution is used. Alternately, some 1:100 dilution steps can be used to replace two 1:10 dilutions. For example, you could achieve the same dilution by using two bottles of 99.0 ml of water or saline (0.9%) and three tubes with 9.0 ml of water or saline (0.9%).
2. Nutrient agar deeps (18.0 ml). You will need one for each pour plate. Typically, you would plate out three dilutions that would bracket the expected number of colonies and produce at least one dilution with between 30 and 300 colonies. If you do not know about how many colonies to expect, you should plate out more dilutions. You will need at least three and as many as seven deeps. (If triplicate plating is required, multiply these by three.)

    *Note:* Plate counts are often done in *triplicate*. However, for reasons of economy, your laboratory instructor may require only one plating of each dilution. If so, remember that the purpose of this exercise is to demonstrate the technique. In practice, triplicate platings should be considered to be standard procedure.
3. Sterile, 1.0-ml pipettes. Read the section of Chapter IV that discusses pipetting errors. You will need at least one pipette for each dilution step.
4. Sterile petri plates.
5. Sample to be plated. Any available source such as water, soil, milk, or a food product may be used. Eighteen to 24-hour broth cultures of *Staphylococcus epidermidis* or *E. coli*, with populations of about 1 million bacteria/ml may also be used.

### Part Two: Membrane-Filter Counts

1. Membrane-filter apparatus. This should include a receiving funnel or vessel, filter holder, and suction funnel. Complete assemblies are available commercially. These usually include all components, including the filter, in a single assembly.
2. Vacuum source or pump.
3. Sterile membrane filters, 0.45-$\mu$m pore size, to fit membrane-filter apparatus.
4. Sterile filter-paper absorbent pads, same diameter as filters.
5. Sterile *m*-Endo MF broth (not more than three days old, stored under refrigeration) or other suitable broth medium.
6. Sterile 5.0-ml pipettes.
7. 20 ml of sterile water for rinsing filters.
8. Water samples, including at least one that is known to contain coliforms. Students might alternate samples, so that adjacent groups test coliform-free and contaminated samples.

## LABORATORY PROCEDURES

### Part One: Pour-Plate Procedures

1. Obtain and label all of the dilution blanks, nutrient agar deeps, pipettes, and petri plates you will need to complete your dilutions and platings. Remember to label the petri plates on the top (bigger half) so that your labels will not obstruct your view through the bottom of the plate.
2. Maintain the agar deeps in a water bath (at about 50°C) to prevent the medium from solidifying before you use it.
3. Arrange your material so that everything is within reach.
4. Decide which of the dilutions will be plated out. If the culture or sample you are working with has about 10,000,000 cells/ml, you should at least plate out the 1:100,000 through the 10,000,000 dilutions. If necessary, ask your instructor for help.
5. Following the diagram in Fig. E24-1, complete your serial dilutions and plating simultaneously. Use the same pipette to transfer 1 ml from one dilution blank to the melted medium *and* to the next dilution blank.

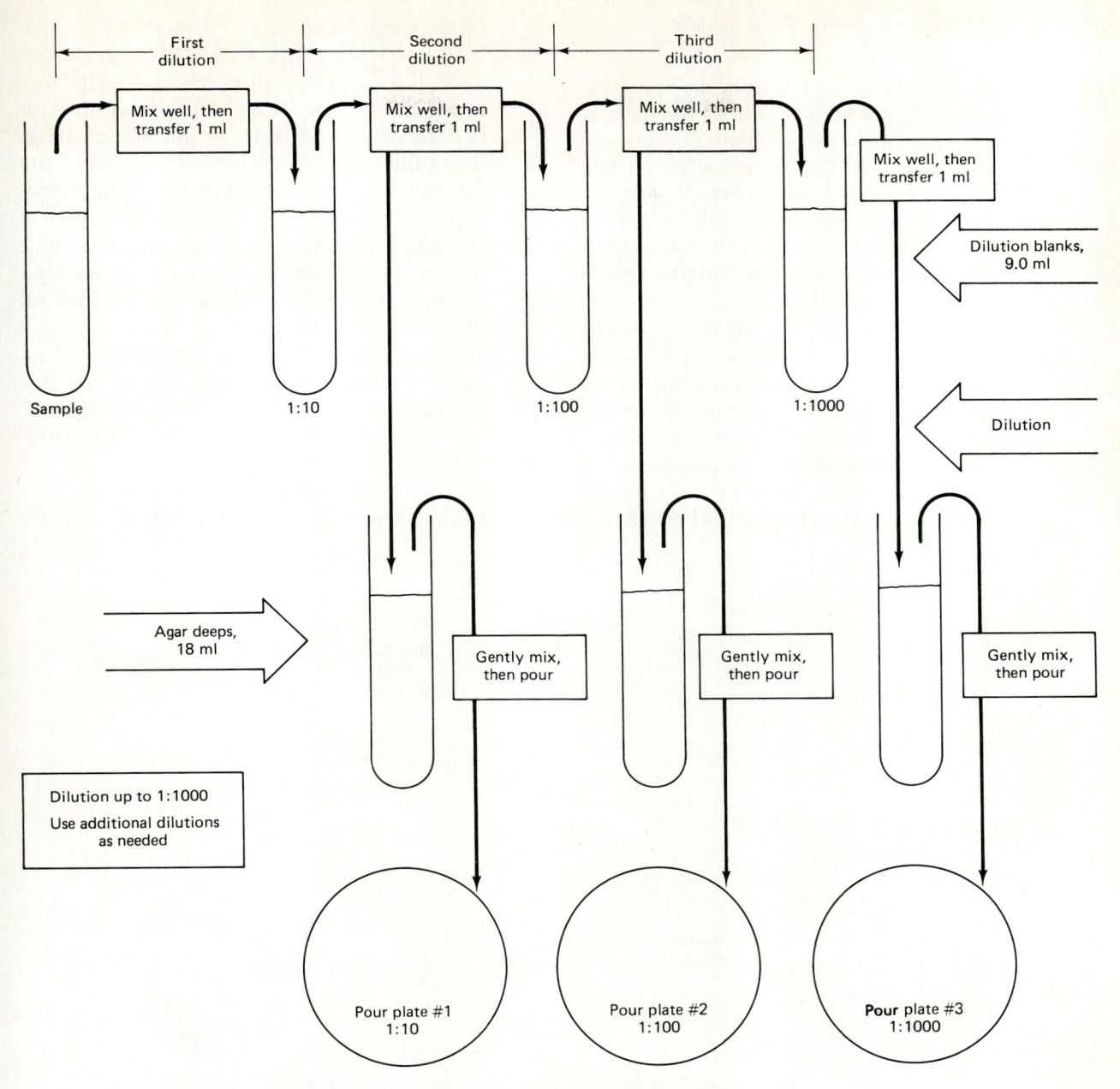

**Figure E24-1** Serial-dilution protocol.

6. Allow the poured medium to solidify before moving the plates.
7. Incubate the plates for 24 hours.
8. Count the number of colonies on each plate. If the medium is transparent, you should count the colonies through the bottom of the plates. It is sometimes helpful to mark each colony with a felt-tip pen as you count it.
9. Use any plates that have between 30 and 300 colonies to calculate the number of colony-forming units (CFU) per milliliter in the original sample. Don't forget to take into account your dilution steps. You do not need to allow for the dilution of the sample by the medium. (Why not?)
10. Record all results and calculations in the Laboratory Report Form.

### Part Two: Membrane-Filter Techniques

1. Assemble a membrane-filtration apparatus. The diagram in Fig. 24-2 shows the general arrangement of the components. Clamp the assembly together, being careful that there are no leaks around the filter. If you are using a prepared apparatus, follow the instructions that came with it. *Remember to carefully observe aseptic techniques, especially when handling the filter.*
2. Using aseptic techniques, place a nutrient pad into a petri plate. Add the required amount of nutrient medium—about 2 ml is usually enough. Set the plate aside while you complete the filtration steps.
3. Pour the water sample into the receiving funnel and apply vacuum to the suction flask. After the sample passes through the filter, rinse the sides of the receiving funnel with about 20 ml of sterile water or saline.
4. Turn the vacuum source off and gently remove the vacuum hose. Allow a minute or two for the air pressure in the suction flask to equilibrate. If you leave the vacuum on for too long, you may kill the cells by excessive drying.
5. Carefully remove the clamp and receiving funnel. If necessary, use sterilized forceps to separate the filter from the bottom of the receiving funnel.
6. Using sterilized forceps, transfer the filter into the culture plate, centering it over the saturated nutrient pad. The medium will be drawn up into the filter by capillary action.
7. Close the petri plate (if you are using specially designed plates the cover will fit

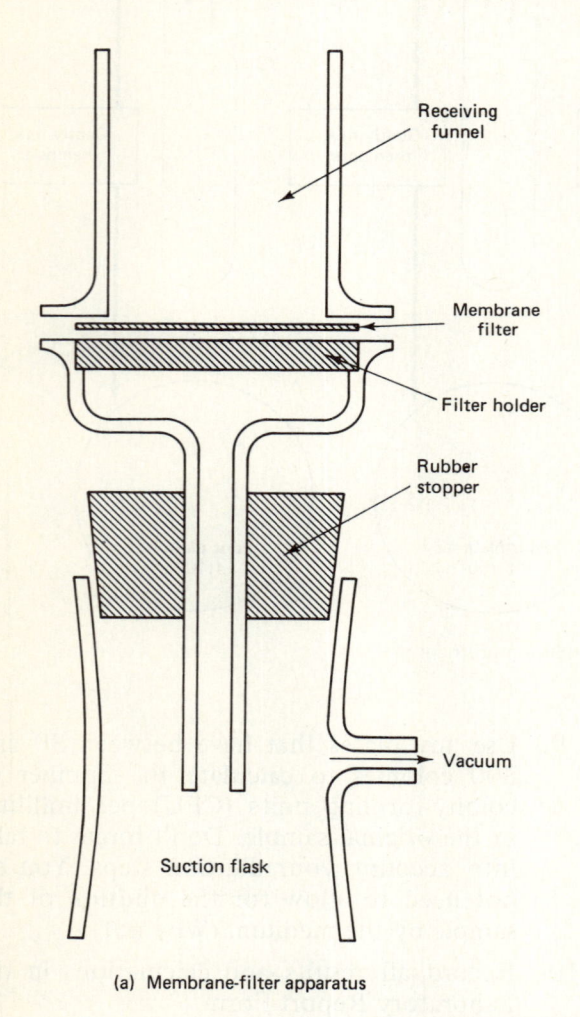

(a) Membrane-filter apparatus

(b) Membrane-filter culture plate

(c) Membrane filter with colonies

(d) Membrane filter in culture plate

**Figure E24-2** Membrane-filter techniques.

tightly), and incubate for 22 to 24 hours. *Do not invert the petri plate.*

8. Count the colonies that appear on the filter. If you used *m*-Endo MF broth, coliforms will produce a distinctive golden to green metallic sheen.

9. Record all results in the Laboratory Report Form.

## REFERENCES

### Text References

1. Atlas, Ronald M., *Microbiology: Fundamentals and Applications.* Chapter 9.
2. Brock, Thomas D., David W. Smith, and Michael T. Madigan, *Biology of Microorganisms,* 4th ed. Chapter 7.
3. Jensen, Marcus, *Introduction to Medical Microbiology.* Chapters 6 and 7.
4. Nester, Eugene W., et al., *Microbiology,* 3rd ed. Chapter 4.
5. Stanier, R. Y., E. A. Adelberg, and J. Ingraham, *The Microbial World,* 4th ed. Chapter 5.
6. Stanier, R. Y., et al., *Introduction to the Microbial World.* Chapter 5.
7. Wistreich, George A., and Max D. Lechtman, *Microbiology,* 4th ed. Chapter 6.

### Resource References

1. Gerhardt, P., ed., *Manual of Methods for General Bacteriology.* Chapter 11.

# LABORATORY PROTOCOL

## Exercise 24—Bacterial Population Counts: Viable Cell Counts

Check each step when you complete it.

### Part One: Pour-Plate Techniques

_____ 1. Set up a water bath to prevent your agar deeps from solidifying prematurely.

_____ 2. Determine what dilution sequence you will use (either simple tenfold serial dilutions or a combination of 1:100 and 1:10 dilution steps.)

_____ 3. Obtain and label your dilution blanks, nutrient agar deeps, and petri plates. Remember to label the plates on the larger half. Obtain the bacterial sample you are going to count.

_____ 4. Arrange your materials so that they are within reach and ready to go.

*Note:* As you make your dilutions, you should be careful to treat each dilution step the same. If a vortex mixer is available, mix each dilution for the same period of time. Alternately, if you shake the blanks, each should be shaken the same number of times (e.g., 25).

_____ 5. Proceed with your plate count. Use the following general protocol (as shown in Fig. 24-1):

    _____ a. Mix the sample carefully.

    _____ b. Transfer 1 ml of the sample to the first dilution blank.

    _____ c. Discard the pipette in a suitable container. Mix the dilution blank.

    _____ d. At each dilution that requires plating, transfer 1 ml to the next blank, then, using the same pipette, transfer a second milliliter to the melted medium. Discard the pipette in a suitable receptacle.

    _____ e. Immediately mix the medium by gentle rocking and pour it into the appropriate petri plate.

    _____ f. Repeat as needed to complete the plate count.

_____ 6. Incubate all plates for 18 to 24 hours.

_____ 7. Count any colonies that develop on the plates. Record your results in the Laboratory Report Form.

### Part Two: Membrane-Filter Techniques

_____ 8. Assemble a membrane-filter apparatus. Use Fig. 24-2 as a guide and/or use the instructions that are provided with any prepared filtration units.

_____ 9. Assemble the petri-plate culture chamber for the filter. Be careful not to contaminate the absorbent nutrient pad or the medium you add to it.

_____ 10. Pour the water sample into the upper, receiving funnel; connect the apparatus to the vacuum source; and allow the sample to pass through the filter.

_____ 11. Rinse the receiving funnel with about 10 to 20 ml of sterile water or saline, then turn off the vacuum and remove the apparatus from the vacuum source.

_____ 12. Disassemble the apparatus and transfer the filter to the culture plate, carefully positioning the filter over the absorbent pad.

_____ 13. Incubate the culture plate for 18 to 24 hours.

_____ 14. Count any colonies that develop and record the results in the Laboratory Report Form.

# NOTES

Name:_____

## LABORATORY REPORT FORM

**Exercise 24—Bacterial Population Counts: Viable Cell Counts**

1. Report your plate-count results on the following table. Show the actual number of colonies counted at each dilution and calculate the number of CFU/ml in the original sample. Finally, convert the counts to their logarithmic numbers.

### PLATE COUNT RESULTS

| Observed Data | | Calculated Results | |
|---|---|---|---|
| Dilutions Counted | Colonies Counted | CFU/ml Actual Count | CFU/ml Log 10 of Count |
| _____ | _____ | _____ /ml | _____ /ml |
| _____ | _____ | _____ /ml | _____ /ml |
| _____ | _____ | _____ /ml | _____ /ml |
| _____ | _____ | _____ /ml | _____ /ml |
| _____ | _____ | _____ /ml | _____ /ml |
| Mean for all counts: | | _____ /ml | _____ /ml |

2. Report the following data for the sample you counted by membrane filtration.

   Colonies (total) per milliliter: _____
   Colonies (coliform) per milliliter: _____
   Colonies (total) in sample: _____
   Colonies (coliform) in sample: _____

## ANSWER THE FOLLOWING QUESTIONS

1. List three advantages of the membrane-filter technique.

2. What is meant by the term *selective medium* and how can it be applied to both the plate-counting procedure and the membrane-filtration counting?

3. List two circumstances when the standard plate-count procedure would be the method of choice.

4. Why can membrane filters be used to sterilize heat-labile substances?

5. Assume that the culture of *E. coli* has cells with diameters of 1 μm and are 3 μm long, and that *S. epidermidis* consist of cells that are 1 μm in diameter. Based on your counts, calculate the total volume and weight of the bacteria in the sample.

# Exercise 25
# Spectrophotometric Methods

As you are now well aware, bacterial suspensions are turbid; when you inoculate a tube of broth, it becomes increasingly turbid as the culture matures. You may already suspect that the more bacteria present in the suspension, the more turbid the suspension will be.

In this exercise you will attempt to determine, by constructing an absorbance-cell number standard curve, if there is a direct relationship between the number of bacteria per milliliter and the turbidity of the suspension. You will then use the curve to estimate the amount of growth that occurs in a culture over the time of your laboratory period.

## PHOTOMETRY

### Absorption of Light by Bacteria

Bacteria, when suspended in a clear fluid, both scatter and absorb light. The amount of light absorbed or scattered is proportional to the density of the bacterial suspension. In other words, the amount of light absorbed or scattered can be graphically related to the number of bacteria in the suspension. Some modern spectrophotometers do not directly measure the light that is absorbed or scattered, but rather measure the amount of light that actually passes through (is not absorbed or scattered by) the sample.

The relative opacity of the sample is reported as either *percent transmission* (%T) or *optical density* (O.D.). Percent transmission indicates the percent of light that passes through the sample—that is, the light that is neither absorbed nor scattered. Optical density, on the other hand, indicates how much light *is* absorbed or scattered. Transmittance and absorbance measure the same phenomenon, although the measure used in one is from the opposite perspective as the one used in the other. However, because they do measure the same phenomenon, one can be calculated from the other.

$$O.D. = 2 - (\log \text{ of } \%T)$$
or
$$O.D. = \log(100\%/T)$$

### Spectrophotometers

A spectrophotometer is a device that can measure the intensity of light. Study Fig. E25-1. The *light source* produces light over a wide range of wavelengths in either the visible or ultraviolet

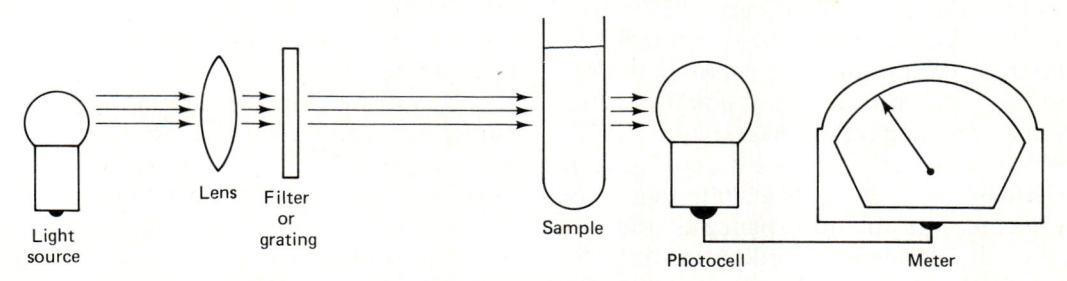

Figure E25-1  Diagram of a single-cell spectrophotometer.

range of the spectrum. Some spectrophotometers have more than one light source that can be switched into or out of the light path, as needed. A lens system is needed to focus the light. The *filter* or *diffraction grating* selects the proper wavelength of light while eliminating others. The light entering the sample will be, for all practical purposes, monochromatic. This is important because different chemicals absorb different wavelengths of light, and if the wrong wavelength is used, no absorbance will be observed.

The *photoelectric cell* is a special tube that converts light energy into electrical energy. When such a tube is correctly connected to a galvanometer, the amount of electrical energy can be measured and can be shown to be directly proportional to the amount of light striking the photoelectric cell.

To use the spectrophotometer, you should adjust the galvanometer to show a transmittance of 100% or an optical density of 0.00 when a tube containing sterile, clear broth is in the sample holder. Then, as shown in Fig. E25-2, when a tube containing turbid broth (such as a culture of bacteria) is inserted into the sample holder, the galvanometer will register a different reading because some of the light will have been scattered or absorbed by the bacterial cells. Typically, the meter will show a lower percent transmission or a higher optical density.

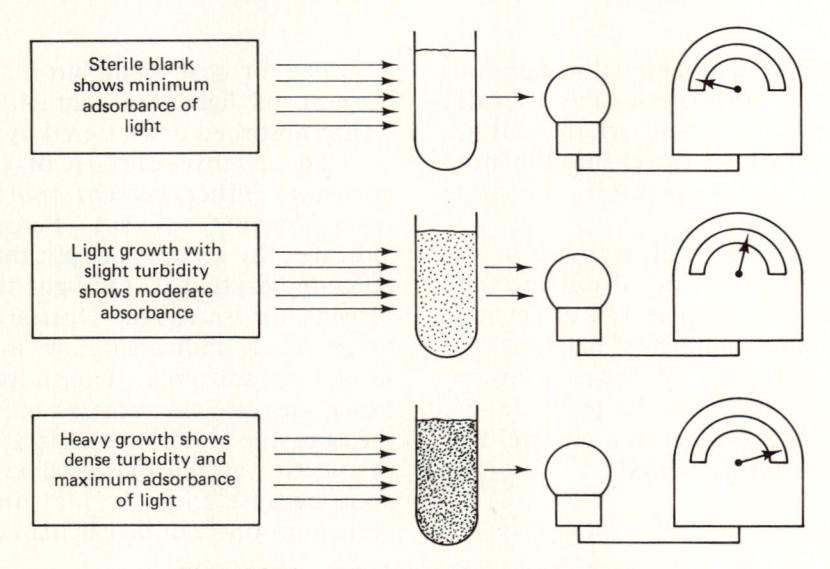

**Figure E25-2** Effect of turbidity on absorbance.

## Standard Curves and How to Use Them

If you were to plot absorbance or percent transmission against the number of bacteria per milliliter of sample, you would, over at least part of the curve, obtain a line that appears to be a straight line. Furthermore, if you kept the general conditions constant (the species of bacteria, composition of medium, etc.) the curve would be reproducible. This type of curve is frequently referred to as a *standard curve*.

Once the graph was drawn, it could be used to determine the number of bacteria per milliliter by simply reading the percent transmission of optical density of the suspension. It is only necessary to measure the absorbance of the suspension, then determine its coordinates from the graph, to find the number of bacteria per milliliter. Many samples can be read rapidly, saving time and material.

Standard curves, such as the one described here, can be used to estimate the density of various bacterial suspensions. For example, they are frequently used to determine the weight of cells at the beginning and end of growth experiments, to complete bacterial growth curves, and for bioassay procedures where the amount of bacterial growth during a specified incubation period must be determined.

In constructing the graph for your standard curve, you must take into account the logarithmic nature of absorbance. A graph that plots O.D. (on the y-axis) against the number of cells (on the x-axis) should be constructed on arithmetically ruled paper, whereas one that plots percent transmittance against cell number should be drawn on semi-logarithmic paper, with the percent transmittance shown on the logarithmic scale.

## OBJECTIVES OF THIS EXERCISE

There is a "direct" relationship between the number of cells in a suspension and the absorbance of light by that suspension. This relationship can be used to estimate the number of cells in the suspension by using a spectrophotometer to measure absorbance accurately and comparing the reading with a previously constructed standard curve. To be able to do this you should:

1. Understand the relationship between the density of the cells in the suspension and the absorbance and/or scattering of light by that suspension.
2. Understand the indirect relationship between percent transmission (%T) and optical density (O.D.).
3. Understand the use of standard curves and how once the relationships in numbers 1 and 2 are known, the data can be used to determine the density of cells by simple photometric measurements (instead of plate counts).

## MATERIALS NEEDED FOR THIS EXERCISE

1. Bausch & Lomb Spectronic 20 spectrophotometer.
2. Spectrophotometer tubes or cuvettes.
3. Five tubes, each containing 4.0 ml of sterile nutrient broth.
4. One culture tube containing 8.0 ml of sterile nutrient broth supplemented with 0.1% yeast extract and 1.0% glucose.
5. Dilution blanks, 99.0-ml bottles or 9.0-ml tubes, sufficient to dilute a culture to 1:10 million.
6. A 24-hour culture of *E. coli*. You will need a volume of about 10 ml. The culture must be at maximum turbidity. Use nutrient broth supplemented as shown in number 4.
7. Four nutrient agar deeps, 18 ml.
8. Four sterile petri plates.
9. Sterile pipettes, 1.0 ml.
10. Sterile culture tubes.

## LABORATORY PROCEDURES

Before you begin the experimental procedure, you must calibrate the spectrophotometer and prepare the cuvettes.

### Calibration of the Spectrophotometer

1. Turn on the spectrophotometer and allow it to warm up for at least 30 minutes.
2. Set the wavelength knob (on the top of the instrument) to a wavelength between 550 and 650 nm. Using the zero adjust knob (on the front left), adjust the meter to read exactly zero.
3. Pour 4 ml of sterile nutrient broth into a cuvette. Wipe the bottom half with soft tissue, then insert it into the sample holder. The index on the tube should be aligned with the index mark on the tube holder.
4. Close the cover of the sample holder and adjust the meter to read 100% transmittance by rotating the light-control knob (on the right front). Remove the sample of nutrient broth.
5. If the meter does not return to zero transmittance when the cuvette is removed from the holder, you must repeat steps 2, 3, and 4.

### Care of Cuvettes

1. Cuvettes should be washed before use by rinsing with distilled water. If spots remain, wash with a gentle detergent solution and rinse several times with distilled water.
2. Hold the cuvette only by the upper third of the tube. If you must mark the cuvettes, do so only at the top and only with a wax pencil.
3. Rinse the cuvette with distilled water between each reading. Allow the tube to drain by holding a Kimwipe against the open end while the cuvette is inverted.
4. Wipe the outside of the cuvette with Kimwipes, never with a paper towel.
5. Always insert the cuvette into the sample holder with the index marks on both the cuvette and holder aligned.

## Part One: Construction of a Standard Curve

*Note:* The two parts of this exercise may be completed simultaneously. If you are doing both Parts One and Two in the same laboratory period, you should set up Part Two first and complete Part One while Part Two is incubating. The spectrophotometric readings for Part Two should be taken while you are constructing the standard curve for part one.

1. Obtain five culture tubes containing 4.0 ml of sterile nutrient broth. Label them "1:2" through "1:32," respectively. Also obtain one empty sterile tube; label it "Undiluted."

    *Note:* Bacteria will be able to grow quite well in the nutrient broth even at room temperature. If the absorbance readings are not taken at about the same time as the plate counts are completed, significant error can be introduced. To minimize this error, ice-cold broth should be used to make the dilutions, and the tubes should be held in an ice bath until the experiment is completed.

2. Study the protocol shown in Fig. E25-3.

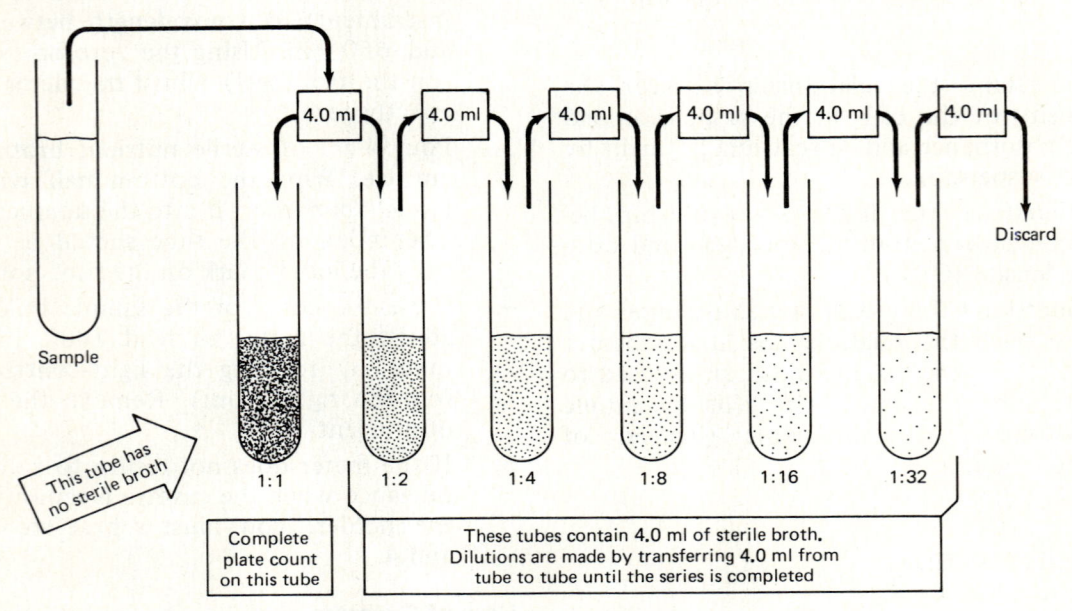

**Figure E25-3** Twofold serial dilution for standard curve (absorbance vs. bacterial count).

3. Transfer 4.0 ml of your sample culture to the tube labeled "Undiluted" *and* to the tube labeled "1:2."

4. Mix the first dilution (1:2) by gently swirling and then transfer 4.0 ml from it to the next dilution tube (the one labeled "1:4"). Mix well.

5. Continue the serial-dilution sequence until all dilutions have been completed (the last one will be the 1:32 dilution.) Discard 4.0 ml from the last tube into a container of disinfectant.

    *Note:* You should determine the optical density of the suspensions in their reverse order. That is, read the most dilute suspension (1:32) first, working your way up to the least dilute (undiluted) suspension. This will minimize error due to carryover of suspension from one sample to another.

6. Confirm that the spectrophotometer has been calibrated. (Refer to the Calibration of the Spectrophotometer section.) Pour the contents of the *last* tube into a cuvette and place the cuvette into the sample holder. Close the cover and record the percent transmittance.

7. After you determine the percent transmission of the sample, pour the broth back into the tube. (Save each sample until the experiment is complete, just in case you have to do it over.) Rinse the cuvette with

distilled water, discarding the rinse water into a container of disinfectant.
8. Repeat steps 6 and 7 until you have measured the percent transmittance of all the tubes, including the undiluted suspension. Blank the spectrophotometer between each reading by ascertaining that a reading of 100% T is obtained with sterile nutrient broth.
9. Record all your readings in the Laboratory Report Form. Calculate the optical density for each reading.
10. Assemble and label the materials needed to complete a plate count. You will need to dilute your sample to 1:10 million and plate out the four highest dilutions.
    a. Dilution blanks (99.0-ml bottles or 9.0-ml tubes).
    b. Agar deeps with 18.0 ml of nutrient agar (melted, held at 50°C).
    c. Sterile petri dishes, appropriately labeled.
    d. Sterile pipettes (one for each dilution step).
11. Review the procedure for plate counting (Exercise 24). Then complete a plate count on the undiluted suspension. Use 1 ml of the first tube ("Undiluted") as your sample.

## Part Two: Measurement of Bacterial Growth by Absorbance

1. Obtain a tube containing about 8.0 ml of sterile supplemented nutrient broth. If possible, use a spectrophotometer cuvette or a new culture tube that shows little if any absorbance with sterile nutrient broth between 550 and 650 nm.
2. Add 1.0 ml of the 24-hour culture of *E. coli* to the sterile broth. Determine the absorbance of this suspension.
3. If this suspension shows a percent transmittance greater than 80%, add an additional amount of the broth culture until the transmittance is less than 80%. Record the reading in the report form.

    *Note:* Step 3 is very important. If the culture is not dense enough, you will be unable to detect changes in the transmittance during the three-hour incubation time. If necessary, you should incubate the culture until its percent transmittance is less than 80%.
4. Incubate the suspension in a water bath at 37°C for the remainder of the laboratory period (or at least three hours), taking the percent transmittance every thirty minutes. Record these readings in the Laboratory Report Form.

## REFERENCES

### Text References

1. Brock, Thomas D., David W. Smith, and Michael T. Madigan, *Biology of Microorganisms,* 4th ed. Chapter 7.
2. Jensen, Marcus, *Introduction to Medical Microbiology.* Chapter 6.
3. Nester, Eugene W., et al., *Microbiology,* 3rd ed. Chapter 4.
4. Wistreich, George A., and Max D. Lechtman, *Microbiology,* 4th ed. Chapter 6.

### Resource References

1. Gerhardt, P., ed., *Manual of Methods for General Bacteriology.* Chapters 3 and 11.

# LABORATORY PROTOCOL

## Exercise 25—Spectrophotometric Methods

Check each step when you complete it.

### Preliminary Procedures

_____ 1. Obtain and clean several cuvettes. Handle the cuvettes by the upper third of the tube.

_____ 2. Set up and calibrate the spectrophotometer.

    _____ a. Allow a 30-minute warmup period.

    _____ b. Set the wavelength between 550 and 650 nm.

    _____ c. Use a cuvette with sterile nutrient broth to set the instrument at 100% transmittance.

*Note:* The protocol is presented in the order the steps should be completed *if you are completing both parts in a single laboratory period.* You should set up the bacterial growth measurement first, then the standard curve. Take the spectrophotometric measurements while you are preparing the standard curve.

Normally, you would have completed the standard curve, then used that data to monitor bacterial growth turbidometrically. Hence the reason for Part One and Part Two.

### Part Two: Use of Photometry to Measure Bacterial Growth

_____ 3. Obtain a culture tube or cuvette containing sterile nutrient broth. Add a sufficient amount of the 24-hour culture of *E. coli* to produce a suspension with a percent transmittance of less than 80%.

_____ 4. Incubate the suspension as directed. You might design an experiment to determine if the growth rate is different at several different temperatures (0, 4, 25, and 37°C).

_____ 5. Measure and record the percent transmittance of the suspension every 30 minutes for the entire laboratory period.

### Part One: The Standard Curve

_____ 6. Obtain five tubes containing 4.0 ml of sterile nutrient broth. Label the tubes "1:2" through "1:32," respectively. Also obtain one empty sterile culture tube and label it "Undiluted."

_____ 7. Transfer 4.0 ml of the bacterial culture to the tube labeled "Undiluted" and to the tube labeled "1:2."

_____ 8. Mix the diluted suspension, then, following the protocol in Fig. 25-3, complete the twofold serial-dilution sequence. Discard 4.0 ml from the last (1:32) tube.

_____ 9. Determine the percent transmittance of each of the suspensions. Start with the most dilute sample, rinsing the cuvettes with distilled water between readings.

**EXERCISE 25**

_____ 10. Obtain and label the following material that you will need to complete a viable plate count.

        _____ a. Dilution blanks (99.0-ml bottles and/or 9.0-ml tubes).

        _____ b. Nutrient agar deeps, 18.0 ml (melted and held at 50°C).

        _____ c. Sterile petri dishes.

        _____ d. Sterile pipettes.

_____ 11. Complete a plate count on the undiluted suspension prepared above. You will need to dilute the suspension to about 1:10 million.

_____ 12. Record these results in the Laboratory Report Form and calculate the optical density of the samples.

# NOTES

Name: _____

# LABORATORY REPORT FORM

## Exercise 25 – Spectrophotometric Methods

1. Count the colonies on all countable plates and calculate the number of colony-forming units per milliliter (CFU/ml) for each dilution, including the undiluted sample.

2. Complete the following table.

### TURBIDITY–ABSORBANCE STANDARD CURVE

| Dilution | CFU/ml | %T | O.D. |
|---|---|---|---|
| Undiluted | | | |
| 1:2 | | | |
| 1:4 | | | |
| 1:8 | | | |
| 1:16 | | | |
| 1:32 | | | |

3. Plot the absorbance/turbidity by graph with the optical density on the y-axis and dilution on the x-axis. Use linearly ruled paper.

Change in absorbance with time

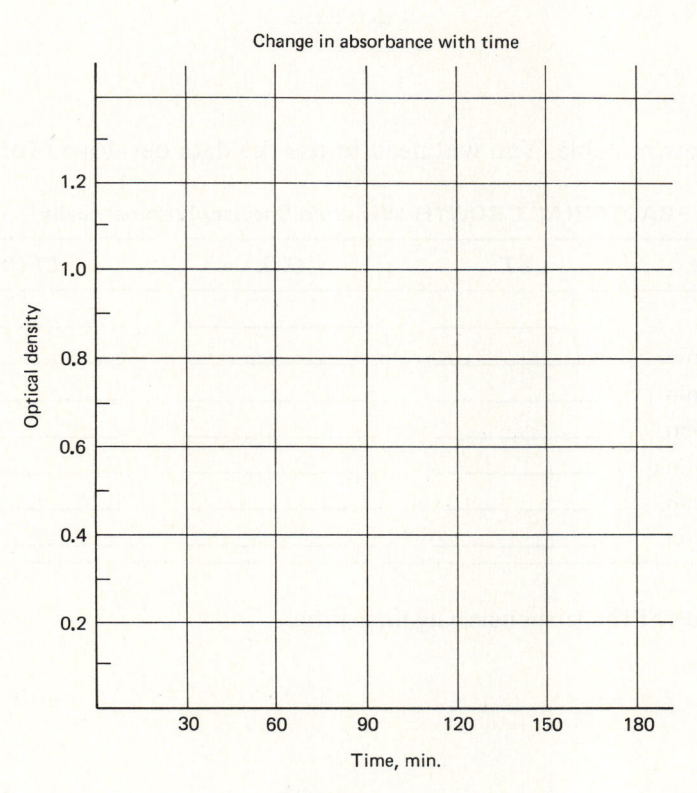

Figure E25-4

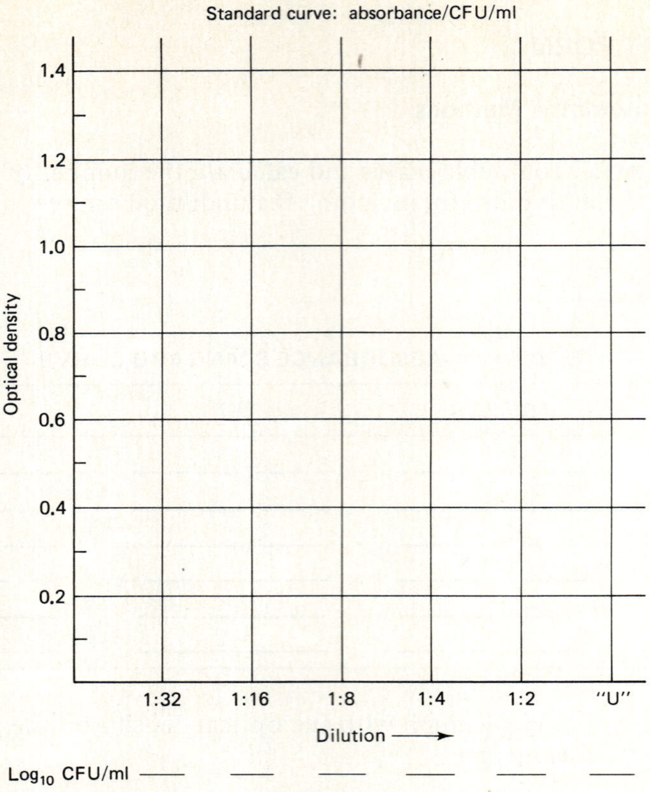
Standard curve: absorbance/CFU/ml
Optical density
1.4
1.2
1.0
0.8
0.6
0.4
0.2
1:32
1:16
1:8
1:4
1:2
"U"
Dilution
Log$_{10}$ CFU/ml

Figure E25-5

4. Complete the following table. You will need to use the data developed for the standard curve.

**BACTERIAL GROWTH (Measured Spectrophotometrically)**

| *Time* | *%T* | *O.D.* | *CFU/ml* |
|---|---|---|---|
| Zero | | | |
| +30 min. | | | |
| +60 min. | | | |
| +90 min. | | | |
| +120 min. | | | |
| +150 min. | | | |
| +180 min. | | | |

5. Plot the absorbance of the suspension against time.

## ANSWER THE FOLLOWING QUESTIONS

1. Explain the relationship between the turbidity of a suspension and the percent transmittance of that suspension.

2. Why can you calculate the optical density of a suspension if you know its percent transmittance? (Or, how are the two measurements related?)

3. List two applications of the principles studied in this exercise.

4. Describe recent applications of turbidity measurements to the determination of the sensitivity of a bacterial isolate to an antibiotic.

# Exercise 26
# Viral Population Counts: Plaque Counting

This exercise, and the two that follow, all involve plaque formation by viruses on permissive hosts. Each exercise examines a different aspect of plaque formation; counting (26), morphology of plaques (27), and host-cell specificity (28). It is possible to combine these exercises or to have separate groups of students in the same lab do one of the three. In any event, to obtain a complete overview of plaque formation, you should read all three of the exercises before completing any one.

A *bacteriophage* is a virus that parasitizes bacteria and, like most viruses, lyses the cell that serves as its host. If the cells that the viruses lyse are immobilized in soft agar, a clear area, or *plaque,* is produced. The plaque appears as a circular, clear zone in an otherwise homogeneously turbid or opaque field. The cleared area results, of course, from the lysis of the bacterial cells in (or on) the agar.

Viruses that infect animal or human cells also destroy their host cells, producing plaques when cultured in tissue-culture systems that use tissue monolayers growing on agar surfaces. However, not all animal cells grow well on agar layers; some must be cultured in rotating glass tubes, with the cells growing on the inner surface of the glass, being constantly bathed with medium. When these tissue-culture systems are used, distinct plaques do not develop. Rather, a characteristic change in cell morphology and physiological activity precedes outright cell death. These changes are often referred to as *cytopathic effect,* or *CPE*.

Plaques are in every way analogous to colonies of bacteria growing on a nutrient medium. The bacteria in the agar are the medium on which the viruses grow. Like a bacterial colony, a plaque can be assumed to represent the progeny of a single phage. The number of plaques can be used to count the number of viruses in a suspension. The titration of a phage suspension is simply the determination of the number of *plaque-forming units (PFU)* present in 1 ml. The most frequently used method for counting phage is to prepare serial dilutions and then to plate the dilutions on lawns of bacterial cells.

## EXPERIMENTAL DESIGN

The standard plate count, as it is used for counting bacteria (see Exercise 24), cannot be used to count viruses. Reproducible plaque counts and consistent plaque morphology require that the bacteria and viruses be suspended together in agar during the time the plaques are developing. The most convenient way of doing this is to employ the agar overlay technique for the actual plaque count.

In the *agar overlay technique,* the virus suspension is mixed with the bacteria in about 3–5 ml of melted agar. The agar is gently mixed and then poured over an agar layer that has already solidified in petri plates. The lower layer of agar provides a nutrient base for the bacterial and viral growth occurring in the upper layer. Restricting plaque formation to a relatively thin surface layer ensures vigorous virus and bacterial growth and clear, easily seen plaques.

In this exercise you will titrate at least one virus suspension. Your instructor may have you do more than one titration so you may observe different types of virus plaque morphologies. Alternately, a different combination of host cell and virus may be assigned to your laboratory partner or to alternating students. In any case, the laboratory procedure will be the same. Study the flow chart shown in Fig. E26-1 and refer to it as you proceed.

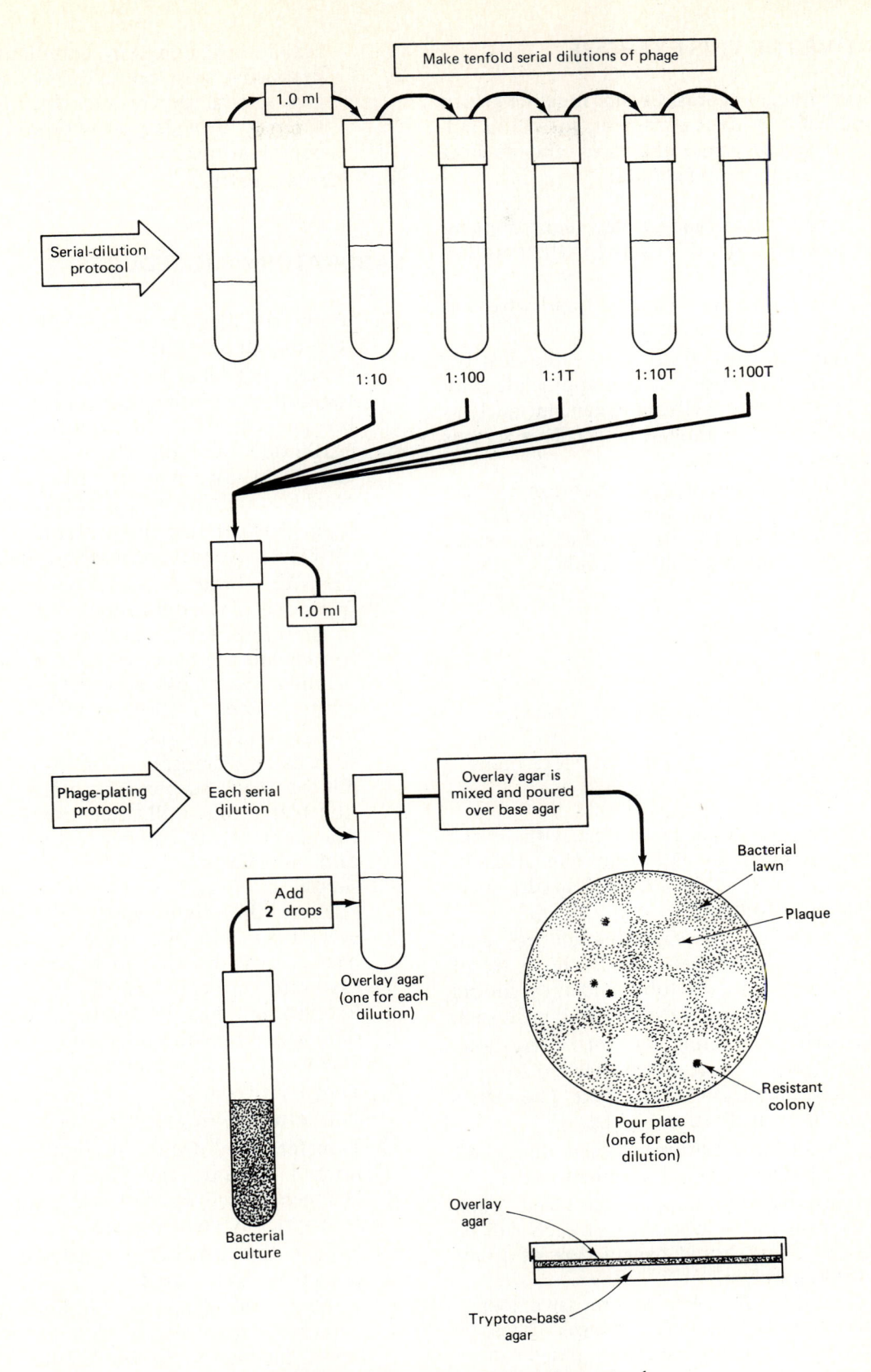

**Figure E26-1** Phage-dilution and -plating protocol.

## OBJECTIVES OF THIS EXERCISE

Viruses are obligate parasites that must be grown on living cells. A virus culture is, in fact, a culture of two organisms: the virus and its host cell. In this exercise you should:

1. Understand the agar overlay technique and the use of serial dilutions for the titration of a virus suspension.
2. Understand what a plaque is and why it is analogous to a virus colony.
3. Understand how viruses cause the destruction of their host cell and appreciate why virus growth in a living organism, such as a human or an animal, is usually a serious pathogenic condition.
4. Understand that plaque formation by bacterial viruses is analogous to plaque formation or cytopathic effect (CPE) by viruses that infect animal or human cells.

## MATERIALS NEEDED FOR THIS EXERCISE

*Note:* These materials will be needed for *each* titration of phage. If you are doing two titrations, you should complete them one at a time.

1. Five prepoured petri plates with about 18 ml of tryptone base agar. The plates of base agar should be warmed in the incubator prior to use.
2. Five tubes of soft tryptone plating agar. There should be between 3 and 5 ml of melted agar in each tube, and they should be kept in a water bath at 50°C until used.
3. Six tubes containing 9.0 ml of tryptone broth. One of these will be needed to wash the host bacterium from a slant. The others will be used as dilution blanks.
4. An 18–24 hour culture of *Escherichia coli* B, or other appropriate permissive host.
5. Suspensions of an appropriate phage (such as T4 or T4r). About 2 ml of each will be needed. They should have a titre of about 10,000 phages/ml.
   *Note: E. coli* is a permissive host for both T4 and T4r phages. These combinations of cell and phage are used in this exercise because they are readily available. Any combination of permissive host cell and its phage will give satisfactory results. Your instructor may provide you with such other combinations.
6. Sterile pipettes.

## LABORATORY PROCEDURES

1. Study the diagram in Fig. E26-1 and refer to it as you proceed.
2. Obtain and label five petri plates and five tubes of soft plating agar. The plates should be labeled with the dilution (1:10 to 1:100,000) and the phage type and host-cell strain. You will also need six dilution blanks containing 9.0 ml of tryptone broth.
3. Keep the soft agar in a water bath at about 50°C to prevent it from solidifying.
4. Transfer about 5 ml of tryptone broth from one of the dilution blanks to the slant of *E. coli* B. Rock the slant back and forth to suspend the bacterial cells. Label the remaining five blanks according to the serial dilution series (1:10 to 1:100,000).
5. Prepare tenfold serial dilutions of the phage suspension. Transfer 1 ml from the undiluted stock suspension to the first dilution blank. Mix well, then transfer 1 ml to the next blank. Repeat these transfers until all dilutions have been made. (There should be five, 1:10 through 1:100,000.)

   At this time check to ensure that everything you will need is ready. The bacteria will be able to tolerate a limited exposure to heat, but the virus will not. Everything must be ready to go so that once you begin the procedure, you will be able to complete it quickly.
6. Transfer two drops of the bacterial suspension into each of the soft agar tubes.
7. Transfer 1 ml from the first phage dilution (1:10) into the first soft agar tube. Mix gently and immediately pour into the appropriately labeled plate.
8. Repeat this procedure until all the dilutions have been plated.
9. Incubate all plates for 24 hours. If you cannot complete the exercise after 24 hours' incubation, store the plates in a refrigerator until the next laboratory period.

10. Complete the Laboratory Report Form for this exercise.

## REFERENCES

### Text References

1. Atlas, Ronald M., *Microbiology: Fundamentals and Applications.* Chapter 9.
2. Brock, Thomas D., David W. Smith, and Michael T. Madigan, *Biology of Microorganisms,* 4th ed. Chapter 10.
3. Jensen, Marcus, *Introduction to Medical Microbiology.* Chapter 31.
4. Nester, Eugene W., et al., *Microbiology,* 3rd ed. Chapter 17.
5. Stanier, R. Y., E. A. Adelberg, and J. Ingraham, *The Microbial World,* 4th ed. Chapter 12.
6. Stanier, R. Y., et al., *Introduction to the Microbial World.* Chapter 7.
7. Wistreich, George A., and Max D. Lechtman, *Microbiology,* 4th ed. Chapter 13.

# LABORATORY PROTOCOL

## Exercise 26 — Viral Population Counts: Plaque Counting

Check each step when you complete it.

### First Day

It is particularly important that you understand each step of this exercise before you begin. There are certain places where you cannot stop to look for material that should be on hand or to get an explanation of the procedure. Read the instructions carefully and ask your instructor to explain any steps you do not understand.

The protocol given here will suggest a sequence in which to complete the exercise. The detailed, step-by-step instructions are given in the text of the exercise.

_____ 1. Assemble all the materials you will need. These should include:

- _____ Petri plates (five) with base agar, warmed prior to use.
- _____ Tubes (five) of soft plating agar, in a water bath at 50°C.
- _____ Dilution blanks (six) of tryptone broth.
- _____ Suspension of phage (T4 or T4r).
- _____ Agar slant culture of a permissive host strain of bacteria (*E. coli* B).
- _____ Sterile pipettes.

_____ 2. Prepare the bacterial suspension.

_____ 3. Prepare the serial dilutions of coliphage.

_____ 4. Add two drops of the bacterial suspension to each tube of soft agar.

_____ 5. Add 1 ml of each of the serially diluted phage suspensions (1:10, 1:100, etc.) to the appropriate soft agar tube. Mix gently and immediately pour over the base agar layer.

_____ 6. Incubate all plates for 24 hours. If plaques cannot be measured after 24 hours' incubation, store the plates in a refrigerator until the next laboratory period.

### Second Day

_____ 7. Count all plaques, measure their diameter, and write a description of the plaque morphologies observed.

_____ 8. Record all data in the Laboratory Report Form and answer all questions.

# NOTES

Name: _____

**LABORATORY REPORT FORM**

**Exercise 26—Viral Population Counts: Plaque Counting**

1. Complete the following table. You should obtain any necessary data from other students in the class and use their plates to determine plaque size and morphology.

RESULTS OF PLAQUE ASSAY

| Phage (Host) | Count PFU/ml | Size, in mm Center | Halo | Morphology and Margin |
|---|---|---|---|---|
| T4 (*E. coli* B) | _____ | _____ | _____ | _____ |
| T4r (*E. coli* B) | _____ | _____ | _____ | _____ |
| (_____) | _____ | _____ | _____ | _____ |

*Note:* The plaque counts must be calculated from the number of plaques that formed on the plates. You should determine the number of plaque-forming units in each milliliter of the stock suspension (PFU/ml).

2. Did you observe any variation in the plaque morphology of any of these phages? If so, what is the significance of such variation?

3. How does your plaque count compare with the expected count (based on the reported titres of the phage suspensions)? Explain any differences.

4. Did any of the plaques have colonies growing within the cleared area? If so, what is the significance of those colonies?

## ANSWER THE FOLLOWING QUESTIONS

1. Describe the *agar overlay technique*.

2. Compare the number of plaques on each of the dilutions. Do they agree with the dilution factor? (Does the 1:10 plate have ten times as many plaques as the 1:100 plate?)

3. How would you detect phage-resistant mutants? Were any detected in this exercise? How would you confirm the resistance to phage?

4. Explain why plaque formation by bacterial viruses is similar to CPE caused by some animal or human viruses.

# Exercise 27
# Phage-Host Cell Interaction
# Part One: Plaque Morphology

Studies of the phenomena associated with bacterial cell and phage interactions have shown that they are directly analogous to the interactions that occur between animal cells and their viruses. Bacterial virology can serve as a direct model for the study of animal virology.

Phage plaques show a typical plaque morphology (remember colonial morphology?). When plated on suitable permissive host cells, the morphology of the plaque is characteristic of the phage and the strain of host cells they were grown on. Of course, if the bacterial strain is not a suitable host (nonpermissive) for the virus, plaques will not be observed at all. While the most apparent trait is the diameter of the plaque, other distinctive characteristics can also be observed. These may include:

- Sharpness of the plaque edge
- Presence of a halo
- Relative size of the halo
- Clarity of the plaque

## PLAQUE MORPHOLOGY OF T4 AND T4r

Phage T4r is a mutant strain of T4 that produces a dramatically different type of plaque. The *r* stands for *rapid lysis*; phages with this genetic characteristic (the ability to cause rapid cell lysis) produce sharp plaque borders with centers that are clear and free of bacteria. In contrast, T4 will produce a plaque with hazy edges, resembling a halo of reduced turbidity around a partially cleared center. The host strain of bacteria for the T4 family of phages is *E. coli* B.

## EXPERIMENTAL DESIGN

The serial dilution and agar overlay techniques (Exercise 26) can be easily modified to demonstrate variations in plaque morphology and host range specificity. By varying the strain of phage or by using mutant varieties, differences in the shape of the plaque can be observed easily.

The consequences of the interaction of the virus with its host cell will be examined in two parts. Part One (this exercise) will test two strains of the same phage (T4 and T4r) against *E. coli* B to demonstrate differences in plaque morphology. In part Two (Exercise 28), two different, morphologically distinct phages (T2 and $\phi$X174) will be titrated against *E. coli* B and *E. coli* C as a demonstration of host-cell specificity.

The two phages (T4 and T4r) will be plated on a permissive host bacterium at two dilutions of the phages. The dilutions are needed to ensure isolated plaques with distinctive and characteristic morphologies. The experimental procedure is shown in Fig. E27-1.

If too many phages are plated, crowded or confluent plaques will be produced and it will be difficult, perhaps impossible, to accurately study their morphology. Incidentally, if you wanted to isolate phage-resistant mutants of the host bacterium, you would use a suspension of viruses that produced confluent plaques. Any colonies that grew would be resistant to the phage.

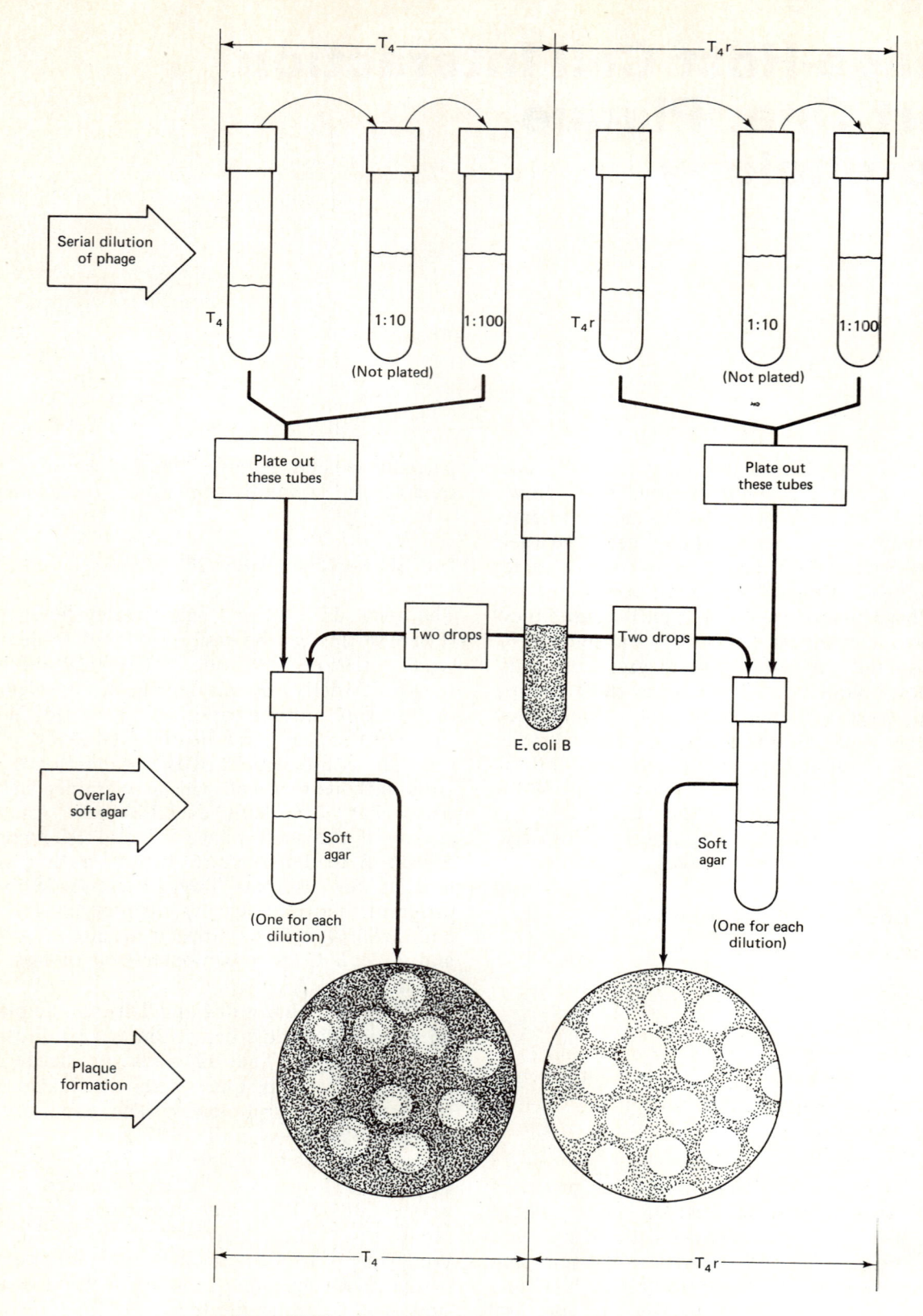

**Figure E27-1** Plaque morphology of *T4* and *T4r* phage.

## OBJECTIVES OF THIS EXERCISE

Viruses are obligate parasites that destroy their host cells and produce plaques when cultured on lawns of suitable host cells. These plaques are often distinctive. In this exercise you should:

1. Learn to distinguish different viruses by their plaque morphology.
2. Understand the nature of rapid lysis, as the term is used in this exercise.
3. Realize the biological consequences of plaque formation for the host cells and understand the clinical significance of host-cell destruction *in vivo*.

## MATERIALS NEEDED FOR THIS EXERCISE

1. Seven prepoured petri plates with about 18 ml of tryptone base agar. Prewarm the plates prior to use.
2. Seven tubes of soft tryptone plating agar. The agar should be melted and the tubes held in a water bath at about 50°C.
3. Five tubes containing 9.0 ml of tryptone broth. One of these will be needed to wash the host bacterium from a slant. The others will be used as dilution blanks.
4. Twenty-four-hour culture of *Escherichia coli* B.
5. Suspensions of T4 and T4r coliphage. About 3 ml of each will be needed. They should have a titre of 10,000 phages/ml.
6. Sterile pipettes.

   *Note:* Prepared kits, such as Carolina Biological BioKit No. 12-4310, are available. If you are using such a kit, most of the materials will be provided in the kit itself. You should follow the instructions that come with the kit, as there may be some variation in procedure. Try to understand any differences that you might observe.

## LABORATORY PROCEDURES

1. Obtain and label seven petri plates and seven tubes of soft plating agar:
   - No. 1: T4—Undiluted
   - No. 2: T4—1:10 (optional)
   - No. 3: T4—1:100
   - No. 4: T4r—Undiluted
   - No. 5: T4r—1:10 (optional)
   - No. 6: T4r—1:100
   - No. 7: Control

   *Note:* The original suspension of phage is the undiluted suspension. You will need to make 1:10 and 1:100 dilutions of the undiluted suspensions.

2. Keep the soft agar in a water bath at about 50°C to prevent it from solidifying.
3. Transfer about 5 ml of tryptone broth from one of the dilution blanks to the slant of *E. coli* B. Rock the slant back and forth to suspend the bacterial cells.
4. Transfer 1 ml of sterile dilution broth to the soft agar tube labeled "Control."

   At this time, check to ensure that everything you will need is ready. The bacteria will be able to tolerate a limited exposure to heat, but the virus will not. Everything must be ready to go so that once you begin the procedure, you will be able to complete it quickly.

   These instructions are given for one of the phages. After steps 8 through 10 have been completed for T4 phage, they should be repeated for the T4r sample.

   You will be preparing a 1:10 and 1:100 dilution of the phage suspension. Counts must be run on the undiluted phage suspension and on the 1:100 dilution (counting the 1:10 dilution is optional).

5. Obtain four dilution blanks and label them:
   - T4—1:10
   - T4—1:100
   - T4r—1:10
   - T4r—1:100

6. Prepare the 1:10 and 1:100 dilutions of both phage suspensions. Transfer 1 ml from the stock to the 1:10 dilution blank, mix it well, then transfer 1 ml of that dilution to the 1:100 blank.
7. Transfer two drops of the bacterial suspension into the soft agar tube labeled "Control" (number 7). Mix the tube gently and immediately pour the contents of the control tube into the control plate (number 7).
8. Transfer two drops of the bacterial suspension into each of two of the tubes of soft agar labeled "T4" (tubes 1, 2, and 3).

Invert the soft agar tubes to mix the suspension.

9. Transfer 1 ml of the original, undiluted T4 phage suspension into the soft agar tube labeled "T4—Undiluted" (number 1). Mix gently and immediately pour into the plate labeled "T4—Undiluted" (number 1).
10. Transfer 1 ml of the 1:10 dilution to the soft agar tube labeled "T4—1:10" (number 2). Mix well and immediately pour into the plate labeled "T4—1:10" (number 2). Repeat this procedure to plate out the 1:100 dilution (tube and plate number 3).
11. Repeat steps 8 through 10 with the T4r phage suspension. Remember that the phage cannot be kept in the melted agar for more than a minute or two. (You will be using plates and tubes number 4, 5, and 6 instead of 1, 2, and 3.)
12. Incubate all plates for 24 hours. If plaques cannot be measured after the 24-hour incubation, store them in a refrigerator until the next laboratory period.
13. Complete the Laboratory Report Form for this exercise.

## REFERENCES

### Text References

1. Atlas, Ronald M., *Microbiology: Fundamentals and Applications.* Chapter 9.
2. Brock, Thomas D., David W. Smith, and Michael T. Madigan, *Biology of Microorganisms,* 4th ed. Chapter 10.
3. Jensen, Marcus, *Introduction to Medical Microbiology.* Chapters 31 and 32.
4. Nester, Eugene W., et al., *Microbiology.* 3rd ed. Chapter 17.
5. Stanier, R. Y., E. A. Adelberg, and J. Ingraham, *The Microbial World,* 4th ed. Chapter 12.
6. Stanier, R. Y., et al., *Introduction to the Microbial World.* Chapter 7.
7. Wistreich, George A., and Max D. Lechtman, *Microbiology,* 4th ed. Chapter 13.

# LABORATORY PROTOCOL

**Exercise 27—Phage-Host Cell Interaction**
**Part One: Plaque Morphology**

It is particularly important that you understand each step of this exercise before you begin. There are certain places where you cannot stop to look for material that should be on hand or to get an explanation of the procedure. Read the instructions carefully and ask your instructor to explain any steps you do not understand.

The protocol given here will suggest a sequence in which to complete the exercise. The detailed, step-by-step instructions are given in the text of the exercise.

Check each step when you complete it.

**First Day**

_____ 1. Assemble the materials you will need for Part One. These should include:

_____ Petri plates (seven) with base agar, prewarmed for several hours in an incubator.

_____ Tubes (seven) of soft plating agar.

_____ Dilution blanks (five) of tryptone broth.

_____ Suspensions of T4 and T4r coliphages.

_____ Agar-slant culture of *E. coli* B.

_____ Sterile pipettes.

*Note:* If you use a prepared kit (such as Carolina Biological Supply BioKit No. 12-4310), all materials are provided. You should follow the instructions supplied with the kit to avoid confusion over the labeling used in the kit and that used in this exercise. Except for the labeling and minor changes in sequence, the two sets of instructions are essentially the same.

_____ 2. Prepare the bacterial suspensions and the soft agar control (steps 3 and 4).

_____ 3. Complete the serial dilutions and plaque assay (count) on phage T4 (steps 5 through 10). Counting the 1:10 dilution is optional.

_____ 4. Repeat steps 5 through 10, but this time use the T4r phage suspension (step 11).

_____ 5. Incubate all plates for 24 hours. If you cannot study the plaques after 24 hours' incubation, store them in the refrigerator until the next laboratory period.

**Second Day**

_____ 6. Count the plaques on all plates.

_____ 7. Write a detailed description of each type of plaque. Pay particular attention to the edges of the plaques.

_____ 8. Record all results on the Laboratory Report Form for this exercise and answer all the questions.

# NOTES

Name: _____

**LABORATORY REPORT FORM**

**Exercise 27—Phage-Host Cell Interaction**
**Part One: Phage Morphology**

1. Complete the following table. You may obtain data from other students in the class and use their plates to determine plaque size and morphology.

**PART ONE: PLAQUE MORPHOLOGY**
*E. coli* **B AND PHAGES T4 AND T4r**

| Phage | Count PFU/ml | Size, in mm Center | Halo | Morphology and Margin |
|---|---|---|---|---|
| T4 | _____ | _____ | _____ | _____ |
| T4r | _____ | _____ | _____ | _____ |

*Note:* The plaque counts must be calculated from the number of plaques that formed on the plates. You should determine the number of plaque-forming units in each milliliter of the stock suspension (PFU/ml).

2. Did you observe any variation (within each strain of phage) in plaque morphology? If so, what is the significance of such variation?

3. How does your plaque count compare with the expected count (based on the titres of the phage suspensions)? Explain any differences.

4. Did any of the plaques have colonies growing within the cleared area? If so, what is the significance of those colonies?

271

## ANSWER THE FOLLOWING QUESTIONS

1. What is *rapid lysis*?

2. Compare the number of plaques on each of the dilutions. Do they agree with the dilution factor? (Does the 1:10 plate have ten times as many plaques as the 1:100 plate?)

3. How would you detect phage-resistant mutants? Were any detected in this exercise? How would you confirm the resistance to phage?

4. Briefly describe a procedure used to demonstrate plaque production by an animal or human virus.

5. Explain why T4 and T4r phages cannot be used to demonstrate host specificity with *E. coli* B and *E. coli* C.

# Exercise 28
# Phage-Host Cell Interaction Part Two: Host-Cell Specificity

One important characteristic of both bacterial and animal viruses is the high specificity that occurs between the virus and its host cell. If a particular cell can serve as a productive host for a virus, it is sometimes referred to as a *permissive host*. Identification protocols for some animal and human viruses often include a determination of which types of host cells may serve as permissive hosts. The unknown virus is tested against a series of cell types to determine which ones will allow productive infection by the virus.

Bacterial host-cell specificity is also used in clinical bacteriology to type (classify) isolates of some pathogenic organisms, such as *Staphylococcus aureus*. This use of phage to type bacteria is referred to as *phage typing*. In this procedure, the bacterial isolate is tested against a selection of phages by placing drops of the phage suspension on a plate that has been covered (by swabbing) with the isolate. If the isolate is susceptible to one or more of the phages, a cleared zone will be observed where the cells were lysed.

Phage typing has epidemiological significance because the technique makes it possible to identify strains of bacteria. For example, phage typing will allow you to ascertain that all cases of a food-poisoning incident were caused by the same strain of *Staphylococcus aureus*.

## OBJECTIVES OF THIS EXERCISE

Viruses are obligate parasites that will grow only on specific, permissive host cells. Phage-host cell specificity is very high, approaching a level typical of enzymatic or serological reactions. In this exercise you should:

1. Understand virus-host range specificity.
2. Understand bacterial phage typing and appreciate its epidemiological applications.
3. Appreciate why some animal viruses infect only certain types of cells (organ specific) of specific host species (host specific).

## HOST SPECIFICITY OF T2 AND ϕX174

T2 phage produces small plaques (about 2-mm diameter) when grown on *E. coli* B. ϕX174 cannot infect (nonpermissive) *E. coli* B, but produces fairly large plaques (about 10-mm diameter) on (permissive) *E. coli* C. The distinctly different plaques can be easily recognized when the two phages are used in the same experiment (just in case you accidentally mislabel the tubes or plates). These two phages can also be used to demonstrate plaque morphology, but unlike the phages used in Exercise 24, where different strains of the same phage were used (T4 and T4r), these are different phages, with different host-cell specificities.

In this exercise, you will attempt to infect *E. coli* B with phage ϕX174 and with two dilutions of phage T2. The T2 phage will produce normal plaques, but since ϕX174 cannot infect this strain of bacteria, no plaques will form.

It is necessary to culture ϕX174 on a susceptible host to demonstrate that the phage is

**274** EXERCISE 28

viable (positive control). When *E. coli* C is used as the control host, the phage produces its typically large and distinct plaques. Two additional controls may be used; T2 phage may be tested against *E. coli* C, and *E. coli* B should be plated out to test its viability. These additional controls are explained in step 11 in Laboratory Procedures.

Study the protocol shown in Fig. E28-1. Note that you will attempt to infect *E. coli* B with both phages to demonstrate the host-specificity requirements of this virus.

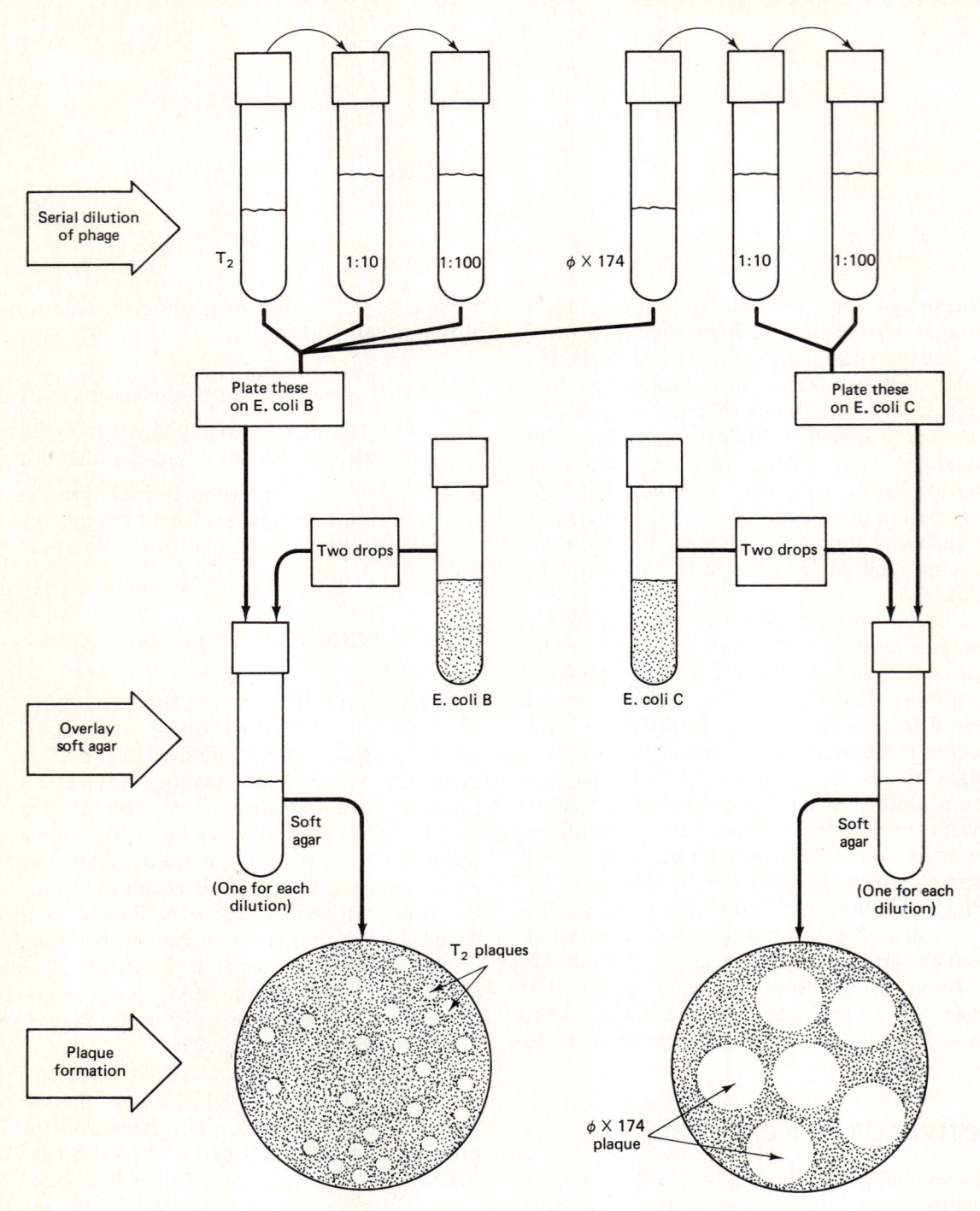

**Figure E28-1** Host specificity and plaque morphology of T2 and φX174 bacteriophage.

## MATERIALS NEEDED FOR THIS EXERCISE

1. Seven prepoured plates with about 18 ml of tryptone base agar. Two additional plates will be needed if the extra controls are used. Warm the plates in the incubator prior to use.
2. Seven tubes of soft tryptone plating agar. The agar should be melted and the tubes held in a water bath at about 50°C.
3. Five tubes containing 9.0 ml of tryptone broth. One of these will be needed to suspend the bacterial host cells. The others will be used as dilution blanks.
4. Twenty-four-hour cultures of *Escherichia coli* B and *Escherichia coli* C.
5. Suspensions of T2 and φX174 coliphages with a titre of between 1,000 and 10,000 PFU/ml.
6. Sterile pipettes.

*Note:* Commercially prepared kits are available, such as Carolina Biological Supply BioKit No. 12-4315, for the demonstration of host specificity. These kits usually contain all of the material needed to complete the exercise. If you use such a kit, follow the instructions provided with the kit. You should try to understand any differences in procedure that you might observe.

## LABORATORY PROCEDURES

1. Obtain and label seven petri plates and six tubes of soft plating agar:
   No. 1: T2–Undiluted (*E. coli* B)
   No. 2: T2–1:10 (*E. coli* B)
   No. 3: T2–1:100 (*E. coli* B)
   No. 4: φX174–Undiluted (*E. coli* B)
   No. 5: φX174–1:10 (*E. coli* C)
   No. 6: φX174–1:100 (*E. coli* C)
   No. 7: Control (*E. coli* C)
   Additional controls:
   No. 8: T2–Undiluted (*E. coli* C)
   No. 9: Control (*E. coli* B)

   *Note:* The original suspension of phage is the undiluted suspension. You will need to make 1:10 and 1:100 dilutions of both (T2 and φX174) phage suspensions.
2. Keep the soft agar in a water bath at about 50°C to prevent it from solidifying. Prewarm the plates in the incubator before use.
3. Transfer about 4 ml of tryptone broth from one of the dilution blanks into each of the *E. coli* cultures. Rock the slants back and forth to suspend the cells.
4. Transfer 1 ml of the sterile dilution broth to the soft agar tube labeled "Control." (Can you explain why the control tube needs 1 ml of sterile broth?)

   At this time, check to ensure that everything you will need is ready. The bacteria will be able to tolerate a limited exposure to heat, but the virus will not. Everything must be ready to go so that once you begin the procedure, you will be able to complete it quickly.
5. Prepare 1:10 and 1:100 serial dilutions of both phages, using the appropriately labeled dilution blanks.
6. Transfer two drops of the *E. coli* C suspension to the soft agar tube labeled "Control." Mix gently by inverting and immediately pour into the plate labeled "Control."

   Additional controls are suggested in the table at the end of these instructions. Detailed instructions are shown in step 11.

   *Note:* Carefully study the flow diagram in Fig. E28-1 and the summary listed in the table at the end of these instructions. Be certain that you completely understand the procedure before proceeding.
7. Transfer two drops of *E. coli* B to each of tubes 1, 2, 3, and 4.
8. Transfer 1 ml of the appropriate phage dilutions into the soft agar tubes. You will use the undiluted suspension as well as the 1:10 and 1:100 dilutions of T2 (for tubes 1, 2, and 3, respectively) and the undiluted suspension of the φX174 (for tube number 4). Mix gently and pour into the similarly labeled plates. Be careful not to mix things up.

   *Note:* The plating protocol shown in step 8 plates phage T2 on its permissive host at three dilutions (plates and tubes number 1, 2, and 3). It also provides for plating phage φX174 on a nonpermissive host (plate and tube number 4).

9. Transfer two drops of the *E. coli* C suspension to tubes 5 and 6.
10. Transfer 1 ml of the φX174 phage dilutions into tubes 5 and 6. Mix gently and pour into the similarly labeled plates.

    *Note:* In step 10, phage φX174 was plated on its permissive host strain of *E. coli* C (tubes and plates number 5 and 6).
11. Complete the optional controls as directed by your instructor. Number 8 tests the permissiveness of *E. coli* C for phage T2, while number 9 is the *E. coli* B cell control.
12. Incubate all plates for 24 hours. Store plates under refrigeration until the plaques can be studied.
13. Study the morphology of the plaques and measure their diameters. Record all data, including plaque counts, in the Laboratory Report Form for this exercise.

### SUMMARY OF DILUTION AND PLATING PROTOCOL

| | |
|---|---|
| Tube and Plate No. 1: | *E. coli* B and 1.0 ml of undiluted T2 phage suspension |
| Tube and Plate No. 2: | *E. coli* B and 1.0 ml of 1:10 suspension of T2 phage |
| Tube and Plate No. 3: | *E. coli* B and 1.0 ml of 1:100 suspension of T2 phage |
| Tube and Plate No. 4: | *E. coli* B and 1.0 ml of undiluted φX174 phage suspension |
| Tube and Plate No. 5: | *E. coli* C and 1.0 ml of 1:10 suspension of φX174 phage |
| Tube and Plate No. 6: | *E. coli* C and 1.0 ml of 1:100 suspension of φX174 phage |
| Tube and Plate No. 7: | Control—*E. coli* C and 1.0 ml of sterile dilution broth |

*Optional Controls*

| | |
|---|---|
| Tube and Plate No. 8: | *E. coli* C and 1.0 ml of undiluted T2 phage suspension |
| Tube and Plate No. 9: | Control—*E. coli* B and 1.0 ml of sterile dilution broth |

## REFERENCES

### Text References

1. Atlas, Ronald M., *Microbiology: Fundamentals and Applications.* Chapter 9.
2. Brock, Thomas D., David W. Smith, and Michael T. Madigan, *Biology of Microorganisms,* 4th ed. Chapter 10.
3. Jensen, Marcus, *Introduction to Medical Microbiology.* Chapters 31 and 32.
4. Nester, Eugene W., et al., *Microbiology,* 3rd ed. Chapter 17.
5. Stanier, R. Y., E. A. Adelberg, and J. Ingraham, *The Microbial World,* 4th ed. Chapter 12.
6. Stanier, R. Y., et al., *Introduction to the Microbial World.* Chapter 7.
7. Wistreich, George A., and Max D. Lechtman, *Microbiology,* 4th ed. Chapter 13.

# LABORATORY PROTOCOL

## Exercise 28—Phage-Host Cell Interaction
### Part Two: Host Specificity

It is particularly important that you understand each step of this exercise before you begin. There are certain places where you cannot stop to look for material that should be on hand or to get an explanation of the procedure. Read the instructions carefully and ask your instructor to explain any steps you do not understand.

The protocol given here will suggest a sequence in which to complete the exercise. The detailed, step-by-step instructions are given in the text of the exercise.

Check each step when you complete it.

**First Day**

_____ 1. Assemble and label the materials you will need for this exercise. These should include:

_____ Petri plates (seven) with base agar, nine with additional controls. Plates should be prewarmed.

_____ Tubes (seven) of soft plating agar, nine with the additional controls.

_____ Dilution blanks (five) of tryptone broth.

_____ Suspensions of T2 and $\phi$X174 coliphages.

_____ Agar slants of *E. coli* B and *E. coli* C.

_____ Sterile pipettes.

*Note:* If you use the prepared kit (Carolina Biological Supply BioKit No. 12-4315), all materials are supplied. You should follow the instructions provided with the kit to avoid confusion over the labeling used in the kit and that used in this exercise. Except for the labeling and minor changes in procedure, the two sets of instructions are essentially the same.

_____ 2. Prepare the phage dilutions, the bacterial suspensions, and the soft agar control (steps 3 through 6.)

*Note:* Study the flow diagram shown in Fig. 28-1. Be sure that you understand the procedure before proceeding.

_____ 3. Complete the assay of T2 phage and testing of $\phi$X174 phage with *E. coli* B (steps 7 and 8).

_____ 4. Complete the assay of $\phi$X174 phage with *E. coli* C (steps 9 and 10).

_____ 5. Incubate all plates for 24 hours.

**Second Day**

_____ 6. Count all plaques, measure their diameter, and write a description of the different plaque morphologies observed.

_____ 7. Record all data on the Laboratory Report Form and answer all questions.

# NOTES

Name:_____

# LABORATORY REPORT FORM

**Exercise 28—Phage–Host Cell Interaction**
**Part Two: Host-Cell Specificity**

1. Complete the following table. You may obtain data from other students in the class and use their plates to determine plaque size and morphology.

**PART TWO: HOST-CELL SPECIFICITY**
*E. coli* **B AND C AND PHAGES T2 AND** $\phi$**X174**

| Phage (Host) | Count PFU/ml | Size, in mm Center | Halo | Morphology and Margin |
|---|---|---|---|---|
| T2 (*E. coli* B) | _____ | _____ | _____ | _____ |
| $\phi$X174 (*E. coli* B) | _____ | _____ | _____ | _____ |
| $\phi$X174 (*E. coli* C) | _____ | _____ | _____ | _____ |

*Note:* The plaque counts must be calculated from the number of plaques that formed on the plates. You should determine the number of plaque-forming units in each milliliter of the stock suspension (PFU/ml).

2. Were any colonies observed within the plaques? Discuss the significance of such colonies.

3. Did you observe any variations in plaque morphology for phage T2 or $\phi$X174? What is the significance of such variation?

EXERCISE 28

**ANSWER THE FOLLOWING QUESTIONS**

1. How would you determine if the strain of *E. coli* used in the other exercises of this manual was *E. coli* B or C?

2. What is *phage typing*?

3. How would you phage type an unknown strain of *S. aureus* (for example, one that you isolated from a person you thought was suffering from food poisoning)?

4. Describe the epidemiological value of phage typing.

5. What other diagnostic technique(s) show(s) a similar degree of specificity to phage typing?

6. Explain the controls, including the additional controls, used in this exercise.

# Appendix 1
# References

## Text References

1. Atlas, Ronald M. *Microbiology: Fundamentals and Applications.* New York, N.Y.: Macmillan Publishing Co., 1984.
2. Brock, Thomas D., David W. Smith, and Michael T. Madigan. *Biology of Microorganisms,* 4th ed. Englewood Cliffs, N.J.: Prentice-Hall, Inc., 1984.
3. Jensen, Marcus. *Introduction to Medical Microbiology.* Englewood Cliffs, N.J.: Prentice-Hall, Inc., 1984.
4. Nester, Eugene W., et al. *Microbiology,* 3rd ed. Philadelphia, Pa.: Saunders College Publishing, 1983.
5. Stanier, R. Y., E. A. Adelberg, and J. Ingraham. *The Microbial World,* 4th ed. Englewood Cliffs, N.J.: Prentice-Hall, Inc., 1976.
6. Stanier, R. Y., et al. *Introduction to the Microbial World.* Englewood Cliffs, N.J.: Prentice-Hall, Inc., 1979.
7. Wistreich, George A., and Max D. Lechtman. *Microbiology,* 4th ed. New York, N.Y.: Macmillan Publishing Co., 1984.

## Resource References

1. Buchanan, R. E., and N. E. Gibbons, eds. *Bergey's Manual of Determinative Bacteriology,* 8th ed. Baltimore, Md.: Williams and Wilkins Co., 1974.
2. Finegold, Sidney M., and W. J. Martin. *Bailey and Scott's Diagnostic Microbiology,* 6th ed. St. Louis: C. V. Mosby Co., 1982.
3. Gerhardt, P., ed. *Manual of Methods for General Bacteriology.* Washington, D.C.: American Society for Microbiology, 1981.
4. Lennette, E. H., ed. *Manual of Clinical Microbiology,* 3rd ed. Washington, D.C.: American Society for Microbiology, 1980.
5. McGonagle, L. A. *Procedures for Diagnostic Microbiology.* Seattle, Wash.: University of Washington Press, 1978.

# Appendix 2
# Media, Supplies, and Staining Procedures

## MEDIA

Almost all of the media used in this manual may be obtained in dried, powdered form and in prepoured form. In most cases, the expense of the prepoured form is not justified, although convenience and quality may be overriding considerations in some circumstances (see Chapter 2). The author has found it convenient to use the prepoured media in some instances, as indicated by the recommendations shown on the list.

| Medium | Dried Form Recommended | Prepoured Form Recommended |
|---|---|---|
| Bile-esculin agar | + | |
| Bile-esculin agar, with horse serum | | + |
| Blood agar base with 5% sheep RBC | + | + |
| Blood tellurite agar | + | |
| Chocolate agar, with supplement | | + |
| DNAase agar, with methyl green | + | |
| Hektoen enteric agar | + | |
| Lead acetate agar | + | |
| Loeffler coagulated serum slants | | + |
| Lowenstein-Jensen medium | | + |
| MacConkey's agar | + | |
| Mannitol salt agar | + | |
| Martin-Lewis agar | | + |
| Motility medium | + | |
| MR-VP medium | + | |
| Mueller-Hinton agar | + | |
| Nitrate agar (or broth) | + | |
| Nutrient agar | + | |
| Nutrient broth | + | |
| Nutrient gelatin | + | |
| Phenylethyl alcohol (PEA) blood agar | | + |

| Medium | Dried Form Recommended | Prepoured Form Recommended |
|---|---|---|
| Phenol red sugar broths | | |
|    Glucose | + | |
|    Lactose | + | |
|    Mannitol | + | |
|    Sucrose | + | |
| Rogosa SL agar | + | |
| Salmonella-Shigella (SS) agar | + | |
| Selenite F broth | + | |
| SIM agar | + | |
| Simmon's citrate agar | + | |
| Snyder test agar | + | |
| Starch agar | + | |
| Stock culture agar | + | |
| Tetrathionate broth | + | |
| Thioglycolate medium, 1.5% agar | + | |
| Thioglycolate medium, fluid | + | |
| Thiosulfate citrate bile sucrose agar (TCBS) | + | |
| Triple sugar iron (TSI) agar | + | |
| Tryptone broth | + | |
| Trypticase soy agar | + | |
| Trypticase soy broth | + | |
| Trypticase soy broth, with 6.5% sodium chloride | + | |
| Urea agar | + | + |

| Media in Biplates | Prepoured Form Recommended |
|---|---|
| Blood agar/chocolate agar | + |
| Blood agar/MacConkey's agar | + |
| Chocolate agar/Thayer-Martin agar | + |
| PEA-Blood agar/MacConkey's agar | + |

## SPECIAL MEDIA

### Tryptone Soft Agar (for Phage)

| | |
|---|---|
| Tryptone | 10.0 gm |
| Potassium chloride | 5.0 gm |
| Agar | 9.0 gm |
| Distilled water | 1.0 liter |

Dissolve all ingredients in the distilled water; heat to boiling to dissolve the agar. Dispense 5-ml aliquots into small screw-cap tubes. Autoclave for 15 min at 121.5°C.

### Tryptone Base Agar (for Phage)

| | |
|---|---|
| Tryptone | 10.0 gm |
| Calcium chloride, 1.0 M | 2.0 ml |
| Sodium chloride | 5.0 gm |
| Agar | 11.0 gm |
| Distilled water | 1.0 liter |

Dissolve all ingredients in the distilled water; heat to boiling to dissolve the agar. Autoclave for 15 min at 121.5°C. Cool to about 55°C, pour into sterile petri plates. After the agar has solidified, invert and incubate for 24 hours at 37°C.

## Tryptone Dilution Broth (for Phage)

| | |
|---|---|
| Tryptone | 10.0 gm |
| Calcium chloride, 1.0 M | 2.0 ml |
| Potassium chloride | 5.0 gm |
| Distilled water | 1.0 liter |

Dissolve all ingredients in the distilled water. Dispense 9.0-ml aliquots into screw-cap tubes. Autoclave for 15 min at 121.5°C.

## SUPPLIES

### Individual Supplies

One of each recommended for each student.

| | |
|---|---|
| Centimeter ruler | Inoculating needle |
| Felt-tip marker, fine tip | Lead pencil, no. 2 |
| Forceps | Wax marking pencil, water soluble |
| Inoculating loop | |

### General Supplies

Supplies needed to complete the exercises in this manual.

| | |
|---|---|
| Absorbent paper mats | Germicidal hand soap |
| Agglutination plates | Lancettes |
| Agglutination slides | Membrane-filter apparatus, with 0.45-$\mu$ filters |
| Alcohol wipes | |
| Blank paper disks, 6.0-mm diameter | Microscope slides, with frosted ends |
| Breed-counting slides | Ocular micrometers |
| Bunsen burner | Paper towels |
| Counting chambers, blood cell | Paraffin blocks and film |
| | Ring stand |
| Coverslips, 22 mm sq | Spectrophotometer |
| Depression slides, three-well | Spectrophotometer cuvettes |
| Diluting pipettes, re'd and white blood cell | Squeeze bottles, plastic |
| | Stage micrometer |
| Disinfectant, for work area | Staining trays or pans |
| Disposable droppers | Sterile cotton swabs |
| Dropping bottles, stain | Test tube rack, 50-tube |

## PREPARATION OF STAINING REAGENTS

### Stains Needed

| | |
|---|---|
| Basic Fuchsin | Methylene Blue |
| Crystal Violet | Safranin O |
| Malachite Green | Toluidine Blue |

## Acid-Fast Stain, Ziehl-Neelsen Procedure

*Stock solutions:*
Solution A: Dissolve 0.3 gm of basic fuchsin in 10.0 ml of 95% ethyl alcohol.
Solution B: Dissolve 5.0 gm of phenol crystals in 95.0 ml of distilled water.

*Primary stain:* Combine Solution A and Solution B.

*Acid-alcohol rinse:* Mix 3.0 ml of concentrated hydrochloric acid with 97.0 ml of 95% ethyl alcohol.

*Counterstain:* Dissolve 0.3 gm of methylene blue in 100.0 ml of distilled water.

## Capsule Stain, Anthony's Method

*Stain solution:* Dissolve 1.0 gm of crystal violet in 100.0 ml of distilled water.

*Mordant solution:* Dissolve 20.0 gm of copper sulfate in 100.0 ml of distilled water.

## Capsule Stain, Thin-Smear Method

Use good, drawing-quality india ink. The particles of carbon in the india ink should be as small as possible.

## Flagella Stain, Gray's Method

*Stock solutions:*
Solution 1: Dissolve 20.0 gm of tannic acid in 100 ml of distilled water.
Solution 2: Prepare a saturated solution of potassium alum.
Solution 3: Prepare a saturated solution of mercuric chloride.
Solution 4: Dissolve 3.0 gm of basic fuchsin in 100.0 ml of 95% ethyl alcohol.

*Staining Solution A:* Staining Solution A does not store well and it should be prepared fresh daily. Combine the stock solutions in the following proportions: No. 1–2.0 ml, No. 2–5.0 ml, No. 3–2.0 ml, and No. 4–0.4 ml. Filter before use.

*Staining Solution B:* Use Ziehl-Neelsen carbol fuchsin. This is the primary stain used in the acid-fast procedure.

## Gram Stain, Hucker's Method

*Stock solutions:*
- Solution A: Dissolve 2.0 gm of crystal violet in 20.0 ml of 95% ethyl alcohol.
- Solution B: Dissolve 0.8 gm of ammonium oxalate in 80.0 ml of distilled water.

*Primary stain:* Combine Solution A and Solution B. Allow to stand overnight, then filter before use.

*Mordant solution:* Dissolve 1.0 gm of iodine and 2.0 gm of potassium iodide in 300 ml of distilled water. It may be necessary to grind the iodine and potassium iodide in a mortar while adding the water to the mixture.

*Decolorizing solvent:* 95% ethyl alcohol. Some laboratories use a mixture of 100 ml of 95% ethyl alcohol and 100 ml of acetone. Both solvents work, but the alcohol/acetone mixture will discolor the cells more rapidly than the alcohol.

*Counterstain:* Dissolve 0.25 gm of Safranin O in 10 ml of 95% ethyl alcohol, then add the mixture to 100.0 ml of distilled water.

## Metachromatic Granule Stain, Albert's Method

*Primary stain:* Mix 0.2 gm of malachite green and 0.15 gm of toluidine blue in 2.0 ml of 95% ethyl alcohol. Add 100.0 ml of distilled water and 1.0 ml of glacial acetic acid.

*Counterstain solution:* Use Gram's (Lugol's) iodine solution.

## Metachromatic Granule Stain, Loeffler's Method

*Staining solution:* Dissolve 0.3 gm of methylene blue in 30.0 ml of 95% ethyl alcohol, then add 100.0 ml of distilled water. Some laboratories use 0.01% potassium hydroxide instead of distilled water to neutralize acidic impurities occasionally found in the stain.

## Negative Staining

Use good, drawing-quality india ink. The particles of carbon in the india ink should be as small as possible.

## Spore Stain, Schaeffer-Fulton Method

*Primary stain:* Dissolve 5.0 gm of malachite green (oxalate) in 100.0 ml of distilled water. Filter before use.

*Counterstain:* Dissolve 0.25 gm of Safranin O in 10.0 ml of 95% ethyl alcohol, then add 100.0 ml of distilled water.

# Appendix 3
# Chemicals, Prepared Reagents, and Test Procedures

## CHEMICAL REAGENTS

Acetic acid, glacial
Alpha naphthol (Barritt's reagent)
Ammonium oxalate
Bile salts (sodium desoxycholate)
Calcium chloride
Copper sulfate
Dimethyl-alpha-naphthylamine (nitrite test)
Dimethyl-p-phenylenediamine hydrochloride (oxidase test)
Diphenylamine (nitrate test)
Ethyl alcohol, 95%
Formaldehyde
Hydrochloric acid, concentrated
Hydrogen peroxide, 3%
Iodine, crystalline
Mercuric chloride
Methyl red indicator (Difco No. B209)
N-Amyl alcohol (Kovac's indole reagent)
p-Dimethylamine benzaldehyde (Kovac's indole reagent)
Phenol, crystalline
Potassium chloride
Potassium alum (flagella stain)
Potassium hydroxide
Potassium iodide
Silver nitrate
Sodium chloride
Sulfanilic acid (nitrite test)
Sulfuric acid, concentrated
Tannic acid (flagella stain)
Zinc dust
Zinc sulfate

## REAGENT PREPARATION AND TEST PROCEDURES

### Catalase Test

*Reagent preparation:* Ordinary 3% hydrogen peroxide may be used for this test. It is readily available in any drugstore or supermarket.

*Test procedure: Plate-catalase test.* Place one drop of the reagent on the colony to be tested. The appearance of bubbles (oxygen) indicates a positive test. If it is necessary to subculture the colony, it should be done *before* testing or from an identical colony.

*Test procedure: Spot-catalase test.* Place one drop of reagent on the cover of the petri dish or on a slide. Transfer some material from the colony to be tested into the drop of reagent. Bubbling indicates a positive test.

### Coagulase Test

*Reagent preparation:* Commercially prepared coagulase may be purchased from most media and reagent supply houses. It should be reconstituted by adding the appropriate amount of distilled water to the vial (see instructions on label). In an emergency, citrated human plasma may be used, although this is to be considered only as a last resort.

*Test procedure: Slide screening test.* Place two drops of coagulase plasma reagent in the approximate center of a glass slide. Transfer some culture material from the colony to be tested and emulsify it in the drop of plasma. If the organism is coagulase positive, it will agglutinate, whereas a coagulase negative organism will emulsify easily and produce a uniformly hazy suspension.

*Test procedure: Tube incubation method.* Transfer 0.1 ml of the coagulase plasma into a small (12 × 100 mm) tube, then add 0.4 ml of distilled water. Transfer material from several colonies or two to three drops from a 24-hour broth culture into the diluted plasma. Incubate at least four hours. A coagulase-positive organism will coagulate the entire volume of plasma. The mixture may coagulate in a short time, but should not be scored negative until the four-hour incubation period has been completed.

**DNAase Test**

*Reagent preparation:* There are no special reagents required if DNAase agar with methyl green is used.

*Test procedure:* Spot inoculate a DNAase agar plate with the organism(s) to be tested. Bacteria that produce DNAase will cause the dye in the medium to decolorize around the colonies.

**Esculin Hydrolysis (for Enterococci)**

*Reagent preparation:* There are no special reagents required.

*Test procedure:* Inoculate a bile-esculin agar slant with the organism (Enterococcus) to be tested. If the organism hydrolyzes esculin, the medium will turn black. Since this medium contains bile, growth and darkening of the agar represents a presumptive identification of the organism as an Enterococcus.

**Indole Test**

*Reagent preparation:* Combine 75 ml of n-amyl alcohol, 25 ml of concentrated hydrochloric acid, and 5.0 gm of p-Dimethylamine benzaldehyde.

*Test procedure: Medium test.* Place about 0.5 ml of the reagent into the test medium (SIM agar or tryptone broth). A red color will develop in the n-amyl alcohol layer if indol was produced.

*Test procedure: Spot indol test.* Place one or two drops of the reagent on a piece of filter paper. Transfer some material from the colony to be tested onto the wet filter paper. The appearance of a red colony indicates a positive test.

**Methyl-Red Test**

*Reagent preparation:* Dissolve 1.0 gm of methyl red into 300 ml of 95% ethyl alcohol, then add 500 ml of distilled water.

*Test procedure:* Add about 5 drops of methyl red reagent to about 5 ml of a 24-hour MR-VP medium culture of the organism to be tested. A red color indicates a positive test. Note that the color reaction is the opposite of that for phenol red.

**Nitrate Reduction Test**

*Reagent preparation:*
Solution A: Dissolve 8.0 gm of sulfanilic acid in 1 liter of 5 N acetic acid.
Solution B: Dissolve 5.0 gm of dimethyl-alpha-naphthylamine in 1 liter of 5 N acetic acid.

*Test procedure:* This procedure tests for nitrite; bacteria that reduce nitrate to nitrite will give a positive test. Use 24-hour cultures in nitrate broth or nitrate agar.

Add about three drops of Solution A, then three drops of Solution B, to the medium. The formation of a bright red color indicates nitrite is present.

If negative results are obtained (no red color), add a pinch of zinc dust to the medium. If this causes the color to turn red, then nitrate was present and the organism was unable to reduce it. If there is still no color development, it may indicate that all the nitrate was reduced to nitrogen gas or to products other than nitrite. Remember that this is a test for *nitrite*.

Some laboratories include an inner tube (as in the fermentation tests) to trap nitrogen gas.

## Oxidase Test

*Reagent preparation:* Dissolve 1.0 gm of dimethyl-p-phenylenediamine hydrochloride in 100.0 ml of distilled water. The reagent must be prepared fresh daily or stored frozen in 10.0-ml aliquots.

*Test procedure: Spot-oxidase test.* Place a few drops of the reagent on a piece of filter paper. Transfer some culture material to the wet filter paper. A pink, then purple, color indicates a positive test. Nichrome wire can act as a catalyst and may give a false positive test. Best results are obtained with a platinum wire or the tip of a glass Pasteur pipette.

*Test procedure: Plate-oxidase test.* Place one drop of the reagent on the colony to be tested. The colony will turn pink first, then purple, if it is positive. You may choose to flood the plate with reagent (about 2.0 ml) and observe the number of oxidase-positive colonies in a sample (e.g., a throat culture on chocolate agar). The oxidase reagent is toxic to most cells, so transfer to another plate should be completed as soon as the pink color develops.

*Test procedure: Disk test.* Place a Taxo "N" disk close to the colony to be tested. A pink, then purple, color will develop after the reagent has diffused to the colony. Remember that the reagent is toxic and transfer must be from another identical colony, or be completed before the colony becomes dark purple.

## Starch Hydrolysis

*Reagent preparation:* Lugol's iodine (used in the gram-staining procedure) may be used to test for starch hydrolysis.

*Test procedure:* Spot inoculate a starch agar plate with the organism(s) to be tested. After 24 hours of incubation, flood the plate with iodine solution. The starch in the agar will turn purple when exposed to iodine. A clear, nonpurple zone will be observed around colonies of bacteria that hydrolyze starch.

## Voges-Proskauer Test (Acetylmethyl Carbinol, Butylene Glycol)

*Reagent preparation:*
Barritt's
Reagent A:  Dissolve 6.0 gm of alpha naphthol into 100 ml of 95% ethyl alcohol.
Barritt's
Reagent B:  Dissolve 16 gm of potassium hydroxide in 100 ml of distilled water.

*Test procedure:* Transfer about 1 ml of MR-VP medium to a clean test tube. Add about 12 drops of Barritt's Reagent A, then add about 4 drops of Barritt's Reagent B. Mix by gentle agitation. A red color indicates a positive test.

This test requires oxidation of some products of fermentation, and agitation of the mixture will result in faster color development. The test should not be considered negative for at least one half hour after reagents are added.

## PREPARED REAGENTS AND DISKS

### Prepared Reagents

Coagulase plasma

### Reagent Disks

Optochin (bile solubility)      Taxo "N" (oxidase)
Taxo "A" (Bacitracin)           Taxo "P" (bile solubility)

### Antibiotic Disks

Ampicillin, 10 µg              Novobiocin, 30 µg
Chloramphenicol, 30 µg         Penicillin, 10 units
Erythromycin, 15 µg            Polymyxin B, 300 units
Gentamycin, 10 µg              Streptomycin, 10 µg
Kanamycin, 30 µg               Tetracycline, 30 µg
Neomycin, 30 µg

The concentrations of antibiotics given are those used in Exercise 15 and are the correct concentrations for the Kirby-Bauer data given in that exercise. Instructors might want to order one or more of these at different concentrations and invite the students to test the effect of concentration on zone size.

## Antibiotic Solutions

Stock solutions of antibiotics are needed in Exercise 16. They may be made up according to the information contained in the Resource References or in the literature supplied with the antibiotics. For convenience, a stock solution of 1000 µg/ml is recommended. You will need stock solutions of streptomycin and any three other antibiotics.

## Serological Reagents

| | |
|---|---|
| Blood-typing sera: | anti-A, anti-B, and anti-Rh |
| Serotyping reagents: | Polyvalent *Salmonella* O Antiserum Set (e.g., Difco Set: *Salmonella* Poly A-1) |
| | Positive Control Antigen (e.g., Difco *Salmonella* O Antigen, Group B) |
| Febrile agglutination sets: | With positive and negative controls (e.g., Difco Set No. 2407-32-7) |

# Appendix 4
# Cultures Used in This Manual

## SOURCES

Smaller laboratories may find that commercially prepared cultures are often more convenient than keeping stock cultures. This is particularly true of the reference cultures available in dried disk form from Difco and BBL. Carolina Biological Supply has also proven to be a reliable source of cultures. The phage demonstration kits produced by Carolina Biological are highly recommended.

Although some programs may find it more convenient to use commercial cultures, quality control is, and always should be, the responsibility of the user. The author has found that products from the firms mentioned above are generally very reliable.

## BACTERIAL CULTURES

*Bacillus cereus*
*Bacillus megaterium*
*Bacillus subtilis*
*Branhamella catarrhalis*
*Clostridium sporogenes*
*Clostridium tetani*
*Corynebacterium diphtheriae*
*Corynebacterium xerosis*
*Enterobacter aerogenes*
*Escherichia coli*
*Klebsiella pneumoniae*
*Lactobacillus sp.*
*Mycobacterium smegmatis* (or *phlei*)
*Neisseria sicca*
*Neisseria gonorrhoeae*
*Neisseria meningitidis*
*Proteus mirabilis*
*Proteus vulgaris*
*Pseudomonas aeruginosa*
*Salmonella typhimurium*
*Serratia marcescens*
*Shigella flexneri*
*Staphylococcus aureus*
*Staphylococcus epidermidis*
*Streptococcus sp.*
*Streptococcus faecalis*
*Streptococcus pneumoniae*
*Streptococcus pyogenes*
*Vibrio parahaemolyticus*

## PHAGE CULTURES

Coliphage T2
Coliphage T4
Coliphage T4r
Coliphage $\phi$X174
Host Bacteria:
  E. coli B
  E. coli C

## YEAST CULTURES

*Saccharomyces cerevisiae*

## STOCK CULTURES

Stock cultures of these organisms should be maintained according to the procedures recommended in the Resource References. Stock culture agar (Difco) stab cultures held at room temperature work quite well. The organisms should be recultured on a monthly basis.

If dried disk cultures (Difco or BBL) are used, they should be reconstituted as needed, streaked to verify purity, and maintained as stab cultures in stock culture agar.

# Appendix 5
# Bacterial Characteristics Form

The form that follows should be used with the Laboratory Report Forms. The data for it will be accumulated over the first eight exercises.

Use one form for each organism, adding data as you obtain it.

In some cases, you will need to consult the Resource References to look up some of the data to fill in the form.

Name: _____

# BACTERIAL CHARACTERISTICS FORM

**(Complete One Form for Each Organism)**

| NOTES | RESULTS |
|---|---|

Organism: _____

*Cellular Morphology and Staining Reactions*
    Cell shape: _____
    (rod, coccus, vibrio, diphtheroid)
    Gram stain reaction: _____
    Endospore formation: _____
    Acid-fast reaction: _____
    Capsule formation: _____
    Flagella: _____
    Metachromatic granules: _____
    Other morphological features:

*Colonial Morphology*
    Colony color: _____
    Color of agar around colony: _____
    Shape of colony:
    Colony margin:

*Carbohydrate Reactions*
    Glucose fermentation: _____
    Lactose fermentation: _____
    Sucrose fermentation: _____
    Mannitol fermentation: _____
    Hydrolysis of starch: _____
    Methyl red test: _____
    Voges-Proskauer test: _____
    Utilization of citrate: _____
    Thioglycolate growth: _____
    (Aerobic/Anaerobic)

*Protein and Other Reactions*
    Hydrolysis of gelatin: _____
    Indole production: _____
    Hydrogen sulfide production: _____
    Catalase reaction: _____
    Oxidase reaction: _____
    Coagulase test: _____
    Urea hydrolysis: _____
    Nitrate reduction: _____
    DNAase activity: _____
    Motility: _____

Name: _____

## BACTERIAL CHARACTERISTICS FORM

**(Complete One Form for Each Organism)**

| NOTES | RESULTS |
|---|---|
| | Organism: _____ |
| | *Cellular Morphology and Staining Reactions* |
| |     Cell shape: _____ |
| |     (rod, coccus, vibrio, diphtheroid) |
| |     Gram stain reaction: _____ |
| |     Endospore formation: _____ |
| |     Acid-fast reaction: _____ |
| |     Capsule formation: _____ |
| |     Flagella: _____ |
| |     Metachromatic granules: _____ |
| |     Other morphological features: |
| | |
| | *Colonial Morphology* |
| |     Colony color: _____ |
| |     Color of agar around colony: _____ |
| |     Shape of colony: |
| |     Colony margin: |
| | |
| | *Carbohydrate Reactions* |
| |     Glucose fermentation: _____ |
| |     Lactose fermentation: _____ |
| |     Sucrose fermentation: _____ |
| |     Mannitol fermentation: _____ |
| |     Hydrolysis of starch: _____ |
| |     Methyl red test: _____ |
| |     Voges-Proskauer test: _____ |
| |     Utilization of citrate: _____ |
| |     Thioglycolate growth: _____ |
| |     (Aerobic/Anaerobic) |
| | |
| | *Protein and Other Reactions* |
| |     Hydrolysis of gelatin: _____ |
| |     Indole production: _____ |
| |     Hydrogen sulfide production: _____ |
| |     Catalase reaction: _____ |
| |     Oxidase reaction: _____ |
| |     Coagulase test: _____ |
| |     Urea hydrolysis: _____ |
| |     Nitrate reduction: _____ |
| |     DNAase activity: _____ |
| |     Motility: _____ |

Name: _____

## BACTERIAL CHARACTERISTICS FORM

(Complete One Form for Each Organism)

| NOTES | RESULTS |
|---|---|
|  | Organism: _____ |

*Cellular Morphology and Staining Reactions*
- Cell shape: _____
  (rod, coccus, vibrio, diphtheroid)
- Gram stain reaction: _____
- Endospore formation: _____
- Acid-fast reaction: _____
- Capsule formation: _____
- Flagella: _____
- Metachromatic granules: _____
- Other morphological features:

*Colonial Morphology*
- Colony color: _____
- Color of agar around colony: _____
- Shape of colony:
- Colony margin:

*Carbohydrate Reactions*
- Glucose fermentation: _____
- Lactose fermentation: _____
- Sucrose fermentation: _____
- Mannitol fermentation: _____
- Hydrolysis of starch: _____
- Methyl red test: _____
- Voges-Proskauer test: _____
- Utilization of citrate: _____
- Thioglycolate growth: _____
  (Aerobic/Anaerobic)

*Protein and Other Reactions*
- Hydrolysis of gelatin: _____
- Indole production: _____
- Hydrogen sulfide production: _____
- Catalase reaction: _____
- Oxidase reaction: _____
- Coagulase test: _____
- Urea hydrolysis: _____
- Nitrate reduction: _____
- DNAase activity: _____
- Motility: _____

Name: _____

# BACTERIAL CHARACTERISTICS FORM

(Complete One Form for Each Organism)

| NOTES | RESULTS |
|---|---|
|  | Organism: _____ |

*Cellular Morphology and Staining Reactions*
- Cell shape: _____
- (rod, coccus, vibrio, diphtheroid)
- Gram stain reaction: _____
- Endospore formation: _____
- Acid-fast reaction: _____
- Capsule formation: _____
- Flagella: _____
- Metachromatic granules: _____
- Other morphological features:

*Colonial Morphology*
- Colony color: _____
- Color of agar around colony: _____
- Shape of colony:
- Colony margin:

*Carbohydrate Reactions*
- Glucose fermentation: _____
- Lactose fermentation: _____
- Sucrose fermentation: _____
- Mannitol fermentation: _____
- Hydrolysis of starch: _____
- Methyl red test: _____
- Voges-Proskauer test: _____
- Utilization of citrate: _____
- Thioglycolate growth: _____

(Aerobic/Anaerobic)

*Protein and Other Reactions*
- Hydrolysis of gelatin: _____
- Indole production: _____
- Hydrogen sulfide production: _____
- Catalase reaction: _____
- Oxidase reaction: _____
- Coagulase test: _____
- Urea hydrolysis: _____
- Nitrate reduction: _____
- DNAase activity: _____
- Motility: _____

Name: _____

# BACTERIAL CHARACTERISTICS FORM

**(Complete One Form for Each Organism)**

| NOTES | RESULTS |
|---|---|
| | Organism: _____ |
| | *Cellular Morphology and Staining Reactions* |
| |     Cell shape: _____ |
| |     (rod, coccus, vibrio, diphtheroid) |
| |     Gram stain reaction: _____ |
| |     Endospore formation: _____ |
| |     Acid-fast reaction: _____ |
| |     Capsule formation: _____ |
| |     Flagella: _____ |
| |     Metachromatic granules: _____ |
| |     Other morphological features: |
| | |
| | *Colonial Morphology* |
| |     Colony color: _____ |
| |     Color of agar around colony: _____ |
| |     Shape of colony: |
| |     Colony margin: |
| | |
| | *Carbohydrate Reactions* |
| |     Glucose fermentation: _____ |
| |     Lactose fermentation: _____ |
| |     Sucrose fermentation: _____ |
| |     Mannitol fermentation: _____ |
| |     Hydrolysis of starch: _____ |
| |     Methyl red test: _____ |
| |     Voges-Proskauer test: _____ |
| |     Utilization of citrate: _____ |
| |     Thioglycolate growth: _____ |
| |     (Aerobic/Anaerobic) |
| | *Protein and Other Reactions* |
| |     Hydrolysis of gelatin: _____ |
| |     Indole production: _____ |
| |     Hydrogen sulfide production: _____ |
| |     Catalase reaction: _____ |
| |     Oxidase reaction: _____ |
| |     Coagulase test: _____ |
| |     Urea hydrolysis: _____ |
| |     Nitrate reduction: _____ |
| |     DNAase activity: _____ |
| |     Motility: _____ |

Name: _____

# BACTERIAL CHARACTERISTICS FORM

(Complete One Form for Each Organism)

| NOTES | RESULTS |
|---|---|
|  | Organism: _____ |
|  | *Cellular Morphology and Staining Reactions* |
|  |     Cell shape: _____ |
|  |     (rod, coccus, vibrio, diphtheroid) |
|  |     Gram stain reaction: _____ |
|  |     Endospore formation: _____ |
|  |     Acid-fast reaction: _____ |
|  |     Capsule formation: _____ |
|  |     Flagella: _____ |
|  |     Metachromatic granules: _____ |
|  |     Other morphological features: |
|  | *Colonial Morphology* |
|  |     Colony color: _____ |
|  |     Color of agar around colony: _____ |
|  |     Shape of colony: |
|  |     Colony margin: |
|  | *Carbohydrate Reactions* |
|  |     Glucose fermentation: _____ |
|  |     Lactose fermentation: _____ |
|  |     Sucrose fermentation: _____ |
|  |     Mannitol fermentation: _____ |
|  |     Hydrolysis of starch: _____ |
|  |     Methyl red test: _____ |
|  |     Voges-Proskauer test: _____ |
|  |     Utilization of citrate: _____ |
|  |     Thioglycolate growth: _____ |
|  | (Aerobic/Anaerobic) |
|  | *Protein and Other Reactions* |
|  |     Hydrolysis of gelatin: _____ |
|  |     Indole production: _____ |
|  |     Hydrogen sulfide production: _____ |
|  |     Catalase reaction: _____ |
|  |     Oxidase reaction: _____ |
|  |     Coagulase test: _____ |
|  |     Urea hydrolysis: _____ |
|  |     Nitrate reduction: _____ |
|  |     DNAase activity: _____ |
|  |     Motility: _____ |

Name: _____

## BACTERIAL CHARACTERISTICS FORM

**(Complete One Form for Each Organism)**

| NOTES | RESULTS |
|---|---|
| | Organism: _____ |

*Cellular Morphology and Staining Reactions*
    Cell shape: _____
    (rod, coccus, vibrio, diphtheroid)
    Gram stain reaction: _____
    Endospore formation: _____
    Acid-fast reaction: _____
    Capsule formation: _____
    Flagella: _____
    Metachromatic granules: _____
    Other morphological features:

*Colonial Morphology*
    Colony color: _____
    Color of agar around colony: _____
    Shape of colony:

    Colony margin:

*Carbohydrate Reactions*
    Glucose fermentation: _____
    Lactose fermentation: _____
    Sucrose fermentation: _____
    Mannitol fermentation: _____
    Hydrolysis of starch: _____
    Methyl red test: _____
    Voges-Proskauer test: _____
    Utilization of citrate: _____
    Thioglycolate growth: _____
    (Aerobic/Anaerobic)

*Protein and Other Reactions*
    Hydrolysis of gelatin: _____
    Indole production: _____
    Hydrogen sulfide production: _____
    Catalase reaction: _____
    Oxidase reaction: _____
    Coagulase test: _____
    Urea hydrolysis: _____
    Nitrate reduction: _____
    DNAase activity: _____
    Motility: _____

Name: _____

# BACTERIAL CHARACTERISTICS FORM

**(Complete One Form for Each Organism)**

| NOTES | RESULTS |
|---|---|
|  | Organism: _____ |

*Cellular Morphology and Staining Reactions*
    Cell shape: _____
    (rod, coccus, vibrio, diphtheroid)
    Gram stain reaction: _____
    Endospore formation: _____
    Acid-fast reaction: _____
    Capsule formation: _____
    Flagella: _____
    Metachromatic granules: _____
    Other morphological features:

*Colonial Morphology*
    Colony color: _____
    Color of agar around colony: _____
    Shape of colony:
    Colony margin:

*Carbohydrate Reactions*
    Glucose fermentation: _____
    Lactose fermentation: _____
    Sucrose fermentation: _____
    Mannitol fermentation: _____
    Hydrolysis of starch: _____
    Methyl red test: _____
    Voges-Proskauer test: _____
    Utilization of citrate: _____
    Thioglycolate growth: _____
    (Aerobic/Anaerobic)

*Protein and Other Reactions*
    Hydrolysis of gelatin: _____
    Indole production: _____
    Hydrogen sulfide production: _____
    Catalase reaction: _____
    Oxidase reaction: _____
    Coagulase test: _____
    Urea hydrolysis: _____
    Nitrate reduction: _____
    DNAase activity: _____
    Motility: _____

Name: _____

# BACTERIAL CHARACTERISTICS FORM

**(Complete One Form for Each Organism)**

| NOTES | RESULTS |
|---|---|
| | Organism: _____ |
| | *Cellular Morphology and Staining Reactions* |
| |     Cell shape: _____ |
| |     (rod, coccus, vibrio, diphtheroid) |
| |     Gram stain reaction: _____ |
| |     Endospore formation: _____ |
| |     Acid-fast reaction: _____ |
| |     Capsule formation: _____ |
| |     Flagella: _____ |
| |     Metachromatic granules: _____ |
| |     Other morphological features: |
| | |
| | *Colonial Morphology* |
| |     Colony color: _____ |
| |     Color of agar around colony: _____ |
| |     Shape of colony: |
| |     Colony margin: |
| | |
| | *Carbohydrate Reactions* |
| |     Glucose fermentation: _____ |
| |     Lactose fermentation: _____ |
| |     Sucrose fermentation: _____ |
| |     Mannitol fermentation: _____ |
| |     Hydrolysis of starch: _____ |
| |     Methyl red test: _____ |
| |     Voges-Proskauer test: _____ |
| |     Utilization of citrate: _____ |
| |     Thioglycolate growth: _____ |
| |     (Aerobic/Anaerobic) |
| | |
| | *Protein and Other Reactions* |
| |     Hydrolysis of gelatin: _____ |
| |     Indole production: _____ |
| |     Hydrogen sulfide production: _____ |
| |     Catalase reaction: _____ |
| |     Oxidase reaction: _____ |
| |     Coagulase test: _____ |
| |     Urea hydrolysis: _____ |
| |     Nitrate reduction: _____ |
| |     DNAase activity: _____ |
| |     Motility: _____ |

Name: _____

# BACTERIAL CHARACTERISTICS FORM

**(Complete One Form for Each Organism)**

| NOTES | RESULTS |
|---|---|
|  | Organism: _____ |
|  | *Cellular Morphology and Staining Reactions* |
|  |    Cell shape: _____ |
|  |    (rod, coccus, vibrio, diphtheroid) |
|  |    Gram stain reaction: _____ |
|  |    Endospore formation: _____ |
|  |    Acid-fast reaction: _____ |
|  |    Capsule formation: _____ |
|  |    Flagella: _____ |
|  |    Metachromatic granules: _____ |
|  |    Other morphological features: |
|  | *Colonial Morphology* |
|  |    Colony color: _____ |
|  |    Color of agar around colony: _____ |
|  |    Shape of colony: |
|  |    Colony margin: |
|  | *Carbohydrate Reactions* |
|  |    Glucose fermentation: _____ |
|  |    Lactose fermentation: _____ |
|  |    Sucrose fermentation: _____ |
|  |    Mannitol fermentation: _____ |
|  |    Hydrolysis of starch: _____ |
|  |    Methyl red test: _____ |
|  |    Voges-Proskauer test: _____ |
|  |    Utilization of citrate: _____ |
|  |    Thioglycolate growth: _____ |
|  |    (Aerobic/Anaerobic) |
|  | *Protein and Other Reactions* |
|  |    Hydrolysis of gelatin: _____ |
|  |    Indole production: _____ |
|  |    Hydrogen sulfide production: _____ |
|  |    Catalase reaction: _____ |
|  |    Oxidase reaction: _____ |
|  |    Coagulase test: _____ |
|  |    Urea hydrolysis: _____ |
|  |    Nitrate reduction: _____ |
|  |    DNAase activity: _____ |
|  |    Motility: _____ |

Name: _____

## BACTERIAL CHARACTERISTICS FORM

(Complete One Form for Each Organism)

| NOTES | RESULTS |
|---|---|
| | Organism: _____ |
| | *Cellular Morphology and Staining Reactions* |
| |     Cell shape: _____ |
| |     (rod, coccus, vibrio, diphtheroid) |
| |     Gram stain reaction: _____ |
| |     Endospore formation: _____ |
| |     Acid-fast reaction: _____ |
| |     Capsule formation: _____ |
| |     Flagella: _____ |
| |     Metachromatic granules: _____ |
| |     Other morphological features: |
| | |
| | *Colonial Morphology* |
| |     Colony color: _____ |
| |     Color of agar around colony: _____ |
| |     Shape of colony: |
| |     Colony margin: |
| | |
| | *Carbohydrate Reactions* |
| |     Glucose fermentation: _____ |
| |     Lactose fermentation: _____ |
| |     Sucrose fermentation: _____ |
| |     Mannitol fermentation: _____ |
| |     Hydrolysis of starch: _____ |
| |     Methyl red test: _____ |
| |     Voges-Proskauer test: _____ |
| |     Utilization of citrate: _____ |
| |     Thioglycolate growth: _____ |
| |     (Aerobic/Anaerobic) |
| | *Protein and Other Reactions* |
| |     Hydrolysis of gelatin: _____ |
| |     Indole production: _____ |
| |     Hydrogen sulfide production: _____ |
| |     Catalase reaction: _____ |
| |     Oxidase reaction: _____ |
| |     Coagulase test: _____ |
| |     Urea hydrolysis: _____ |
| |     Nitrate reduction: _____ |
| |     DNAase activity: _____ |
| |     Motility: _____ |

# Appendix 6
# The Microscope: Its Use and Routine Care

## ROUTINE CARE OF THE MICROSCOPE

A compound light microscope of reasonably good quality will cost more than a thousand dollars. They are precision instruments and do require reasonable care. If you are not familiar with the parts of a microscope you should read Chapter 1 before continuing. The following rules should be considered mandatory:

**Carrying the microscope.** The microscope is *never carried with one hand*. When you transport the microscope from its cabinet to your work area, one hand supports the microscope from below the base, while the other hand grasps the body.

**Placement of the microscope.** The ocular lenses are usually held in place by gravity. *Do not tilt the microscope* in any way beyond that amount of tilt designed into inclined microscopes. If the scope is too high, raise your stool. Never place the microscope on the edge of the lab bench. Place it as far back from the edge as possible. (How far can you comfortably lean over the lab bench?)

**Peeking and poking.** Perhaps the greatest enemy of an otherwise good microscope is dust and dirt. Every time you remove a lens or any other part of the microscope, dust can enter the light path. You must resist any temptation to look inside the microscope because when you open the scope, the inner surfaces of lenses and prisms can become dirty and dusty. Microscopes are very difficult to clean. If you have a problem, ask your instructor for help—*do not try to fix the problem yourself.*

**Cleaning of lenses.** *Always use lens paper for cleaning lenses.* Do not use any kind of tissue, paper toweling, or cloth. The lenses should be cleaned while they are attached to the microscope, by wiping lightly with a circular motion. The circular motion has the effect of moving particles outward to the edge of the lens, where they are removed. Use distilled water only if you are unable to clean the lens by wiping it with a dry piece of lens paper. If the wet lens paper does not clean the lenses, ask the instructor if you should use lens-cleaning solvent. Clean all lenses—objective, ocular, condenser, and any in the light source. Avoid using the same part of the lens tissue for more than one lens. Always clean the lenses before and after you use the microscope.

**Storage.** The microscope should always be stored in the microscope cabinet. This provides security and allows storage in a relatively dust-free area. Always use the dust cover when one is provided.

**"Your microscope."** You should get into the habit of using the same microscope each time. You will become familiar with that instrument and any of its idiosyncrasies. Any skilled technician is much more comfortable with his or her own tools.

## USE OF THE MICROSCOPE

1. Position the microscope in front of you where you can comfortably look into the ocular lens (lenses).
2. Turn on the light source and position the low-power objective in the light path.
3. Move the substage condenser all the way up, then back it down about one-quarter inch. Be sure that any iris diaphragms are completely open.
4. Look into the ocular lens (lenses) and position yourself at the point where you see a full field of view. You may need to move your head back and forth relative to the lenses to locate this position.

### If You Have a Binocular Microscope:

a. Adjust the ocular lenses to match the distance between your eyes. As you look into the lenses, move them (they move laterally) until you see a single round field of view. Move them back and forth so you will recognize when they are not in adjustment. Remember, adjust the distance between the lenses so that you see a single round field of view. (If you have ever used binoculars you are familiar with the need to do this.)
b. The binocular head of the microscope contains prisms that split the light and compensate for changes in the light path that are necessitated by the adjustments for eye width.
c. The lenses of the binocular head must also be adjusted to compensate for focusing differences between each of your eyes. Notice that one lens is adjustable and one is not. When you have a slide on the microscope, you should focus it as sharply as possible as you look through the nonadjustable lens.
d. After the subject is in sharp focus with the nonadjustable lens, turn the adjustable lens until the image is sharp in both lenses. Now, when you use both objective lenses, the image should be in sharp focus for both eyes. This completes the adjustment of binocular microscopes.
e. Note that graduations are embossed on the side of the adjustable lens and on the body of the binocular head. Make a note of these in your laboratory manual so you can immediately set them each time you use this microscope.

Microscope serial number: _____

Graduation on lens: _____

Graduation on body: _____

5. Place a prepared slide on the microscope stage and center the slide over the hole in the stage. Use the clips provided to hold the slide in place.
6. Using the coarse adjustment, position the low-power objective about one-half inch above the slide.
7. As you look through the microscope, focus on the specimen, using the coarse and fine adjustments to bring the image into sharp focus.

*A word of caution:* Most, but not all, modern microscopes are designed so that it is not possible to touch the slide with the lens. (This is sometimes called *autostop.*) You should develop the habit of positioning the lens so that any focusing adjustments will move the lens away from the slide. This takes practice, but it is well worth the time spent. Whenever possible, use only the fine adjustment while looking through the microscope. Use the coarse adjustment while watching from the side.

8. If the light is too bright, now is the time to reduce the intensity of the light source. This may be done in one of two ways: Reduce the voltage to the light by turning a small knob on the light source housing, or close an iris diaphragm located on the light-source housing (not on the condenser). It will be necessary to readjust the light intensity when you change lenses.
9. The substage condenser is adjusted by moving it up and down until the sharpest-appearing image is attained. Only very

small adjustments in the position of the condenser should be needed. You may also need to adjust the iris diaphragm on the condenser until the image appears free of glare. It is important to learn the difference between real improvements in resolution and apparent improvements caused by increasing contrast. Experiment with the condenser adjustments and be sure to ask for help from your instructor. It will also be necessary to readjust the condenser when you change lenses.

10. After you are satisfied that you have learned to use the low-power objective, position the high-power objective lens over the specimen. This lens is usually a 40× lens and is sometimes referred to as the "high-dry" lens. Do not use the oil-immersion lens at this time.

*Another word of caution:* Modern microscopes are *parfocal*. This means that they are designed so that when one lens is in focus (e.g., the low-power objective), the other lenses (e.g., the high-dry objective) will be very near to focus when positioned over the specimen and should require only minor adjustment of the focus.

The lenses are also designed so they can be rotated into position without hitting the slide. This means that you do not need to move the lenses away from the stage to rotate another lens into position.

Although most microscopes have these features, never assume this to be the case. Always watch from the side as you rotate the lens holder. In all probability, there will be no problem. Watch anyway and avoid serious damage to the lens or the slide. (Sometimes the slide is thicker than usual, or you may be working with a thick specimen and/or cover glass.)

11. You will need to focus the lens, but only small adjustments should be needed. Use the fine adjustment only. Some adjustments will also be needed to increase the amount of light (open the iris diaphragm in the light source or increase the voltage to the lamp). The substage condenser should also be adjusted (the position of the condenser and the diaphragm opening). As before, you should make adjustments to produce the sharpest possible image, distinguishing between increased resolution and increased contrast.

12. When you have become familiar with high-power microscopy, you will be ready to use the oil-immersion objective. Remember that you do not need to change focus or move the lenses away from the slide if the microscope is designed to be parfocal.

13. Rotate the lens nosepiece so that the high-dry lens is moved out of position, about halfway to the next position. Place one or two drops of immersion oil on the center of the slide. Now rotate the nosepiece so that the oil-immersion objective is centered over the slide. The tip of the lens should be in the oil. If it is not, ask your instructor to check the microscope and show you how to proceed.

*A final word of caution:* The space between the lens and the glass slide decreases with magnification. A typical oil-immersion lens with a magnification of 100× will be in focus when it is between 0.1 and 0.2 mm above the subject. The working distance is very small. *When using the oil-immersion objective never use the coarse adjustment.* If you cannot focus with the fine adjustment, ask for help.

14. As with the high-dry lens, adjustments will be needed to increase the intensity of light and to increase resolution. Slight adjustments of the condenser position and diaphragm will be needed. Remember to use only the fine adjustment for focus.

## WHEN YOU ARE FINISHED WITH THE MICROSCOPE

1. Remove the slide from the microscope. If oil was used, carefully wipe the oil from the slide. Return the slide to its container.
2. Use only lens paper to clean the lenses of the microscope.
3. Remove any oil from the oil-immersion lens. If water is needed, use it sparingly on a piece of lens paper and immediately wipe the lens with dry lens paper.
4. Using a different, oil-free piece of lens paper, wipe the high-dry lens and low-power lens with the lens paper. Wipe the lens of the condenser.
5. Position the low-power objective over the center of the stage; return the microscope to the upright position if needed. Any elec-

tric cords should be wrapped around the base in such a way that the microscope will not be resting on the cord when it is stored. Cover the microscope with the dust cover.

6. Return the microscope to its storage cabinet, carrying it with both hands. Be careful not to hit the cabinet (especially the shelf above) with the ocular lenses.

## HELPFUL HINTS

Rules of thumb that you might find useful:

1. When you change lenses to increase magnification, the condenser must be moved closer (higher) to the objective lens.
2. As you increase magnification, the iris diaphragm in the condenser will need to be adjusted.
3. As you increase magnification, your working distance between the tip of the objective lens and the slide will decrease markedly.
4. As you increase magnification, adjustment of the microscope will become more critical.
5. As you use the microscope, try to remember the relationship between resolution, substage condenser, and light source. Try to understand why these rules of thumb apply.